Maintaining Safe Mobility in an Aging Society

Taylor & Francis Group, LLC

From CRC Press - A Taylor & Francis Company
(http://www.crcpress.com)

Book Series on

Human Factors in Transportation

Series Editor

Barry H. Kantowitz
Industrial and Operations Engineering
University of Michigan

Published Titles:

- **Aircrew Training and Assessment**
  Harold F. O'Neil and Dee H. Andrews

- **Automation and Human Performance: Theory and Applications**
  Raja Parasuraman and Mustapha Mouloua

- **Aviation Automation: The Search for a Human-Centered Approach**
  Charles E. Billings

- **Ergonomics and Safety of Intelligent Driver Interfaces**
  Ian Y. Noy

- **Handbook of Aviation Human Factors**
  Daniel J. Garland, John A. Wise and David V. Hopkin

- **Human Factors in Certification**
  John A. Wise and David V. Hopkin

- **Human Factors in Intelligent Transportation Systems**
  Woodrow Barfield and Thomas A. Dingus

- **Principles and Practice of Aviation Psychology**
  Paula S. Tsang and Michael A. Vidulich

- **Stress, Workload, and Fatigue**
  Peter A. Hancock and Paul A. Desmond

- **Ten Questions about Human Error: A New View of Human Factors and System Safety**
  Sidney W.A. Dekker

Forthcoming Titles:

- **Human Factors in Intelligent Transportation Systems, Second Edition**
  John A. Wise, Daniel J. Garland and David V. Hopkin

- **Maintaining Safe Mobility in an Aging Society**
  David W. Eby, Lisa J. Molnar and Paula S. Kartje

Maintaining Safe Mobility in an Aging Society

David W. Eby
Lisa J. Molnar
Paula S. Kartje

CRC Press
Taylor & Francis Group
Boca Raton London New York

CRC Press is an imprint of the
Taylor & Francis Group, an **informa** business

CRC Press
Taylor & Francis Group
6000 Broken Sound Parkway NW, Suite 300
Boca Raton, FL 33487-2742

First issued in paperback 2019

© 2009 by Taylor & Francis Group, LLC
CRC Press is an imprint of Taylor & Francis Group, an Informa business

No claim to original U.S. Government works

ISBN-13: 978-1-4200-6453-7 (hbk)
ISBN-13: 978-0-367-38612-2 (pbk)

Library of Congress Cataloging-in-Publication Data

Eby, David W.
    Maintaining safe mobility in an aging society / David W. Eby, Lisa J. Molnar, Paula S. Kartje.
        p. cm. -- (Human factors in transportation)
    Includes bibliographical references and index.
    ISBN 978-1-4200-6453-7 (alk. paper)
    1. Older automobile drivers. 2. Traffic accidents. 3. Traffic safety. I. Molnar, Lisa J. II. Kartje, Paula S. III. Title. IV. Series.

HE5620.A24E29 2008
362.6--dc22   2008046599

Visit the Taylor & Francis Web site at
http://www.taylorandfrancis.com

and the CRC Press Web site at
http://www.crcpress.com

Dedicated to Al Gorlin (1913–2008),
one of the "faces" of the Michigan Center for Advancing Safe
Transportation throughout the Lifespan (M-CASTL.org)
and someone who exemplified what it means
to age with vitality, spirit, and grace.

Photograph by Gregory Kostyniuk

Contents

Series Foreword

The domain of transportation is important for both practical and theoretical reasons. All of us use transportation systems as operators, passengers, and consumers. From a scientific viewpoint, the transportation domain offers an opportunity to create and test sophisticated models of human behavior and cognition. This series covers both practical and theoretical aspects of human factors in transportation with an emphasis on their interaction.

The series is intended as a forum for researchers and engineers interested in how people function within transportation systems. All modes of transportation are relevant, and all human factors and ergonomic efforts that have explicit implications for transportation systems fall within the series purview. Analytic efforts are important to link theory and data. The level of analysis can be as small as one person or international in scope. Empirical data can be from a broad range of methodologies, including laboratory research, simulator studies, test tracks, operational tests, field work, design reviews, or surveys. This broad scope is intended to maximize the utility of the series for readers with diverse backgrounds.

I expect the series to be useful for professionals in the disciplines of human factors and ergonomics, transportation engineering, experimental psychology, cognitive science, sociology, and safety engineering. It is intended to appeal to the transportation specialist in industry, government, and academia, as well as to the researcher in need of a test bed for new ideas about the interface between people and complex systems.

This text fulfills the goals of the series by discussing the relationships among safety, aging, and mobility. Given demographic trends in the United States and worldwide, the practical aspects of this book are obvious; there are vital safety implications associated with an aging driver population. Theoretical tools applied to this problem include psychological models of perception, psychomotor skill, and cognition, as well as an adaptation of a hierarchical model of driver skills and control and a driving task analysis. Implications of medical conditions and medications upon driving are considered carefully. Screening and assessment tools are presented and evaluated. The final five chapters are devoted to maintaining safe mobility. This goal starts with education and rehabilitation, followed by technology applications inside the vehicle and on the roadway. Using human factors and ergonomic technology to improve the interface between the driver and the vehicle, as well as interactions between the vehicle and the roadway, is a very active research area that is enhanced by attending to the needs of older drivers. The book ends by discussing driving cessation and alternative mobility options that replace driving. Models of human behavior and cognition are applied to the practical issues shaped by the mobility needs of an aging driver population. Forthcoming books in this series will continue this blend of practical and theoretical perspectives.

Barry H. Kantowitz
Sequim, Washington

Preface

The idea for this book came out of the frequent requests we have received over several years from academic departments at the University of Michigan (U-M), practitioners outside the U-M, and community groups for information on aging and mobility. We realized that there was tremendous interest in the issues surrounding aging and safe mobility, but no single resource that drew together the diverse topics that these issues encompass. We hope that this book will serve as such a resource.

The chapters in the book can be categorized into four parts. The first part (three chapters) presents an overview of aging and mobility issues, including discussions of how society is aging, traffic crash risks, age-related declines in functional abilities, and the driving skills needed for safe driving. The first part also presents a framework for the issues related to aging and safe mobility. The second part (two chapters) discusses the current knowledge on the effects of conditions and medications on driving and crash risk. The third part (three chapters) discusses the issues and practices of driver screening and assessment and concludes with a discussion of driver licensing policy. The final part (five chapters) covers topics related to maintaining safe mobility in older adulthood including education and rehabilitation, vehicles and advanced technology, roadway design, driving retirement, and other community mobility options.

Because the study of transportation spans numerous academic fields and transportation issues influence many aspects of our lives, this book is intended for a multidisciplinary audience. Any practitioner who works with older adults (such as a certified driving rehabilitation specialist, occupational therapist, or physician) will find this book useful. This book will also be useful for undergraduate and graduate courses in gerontology or graduate courses in medicine. Finally, the book will be useful to anyone interested in traffic safety, vehicle manufacturing, or roadway design, as the aging population will impact each of these areas in the coming years.

David W. Eby, Lisa J. Molnar, and Paula S. Kartje
Ann Arbor, Michigan

Acknowledgments

We are indebted to several people who helped in the development of this book. We thank the members of the University of Michigan Transportation Research Institute Social and Behavioral Division, in particular, Jonathon Vivoda, Renée St. Louis, Jeri Stroupe, and Lidia Kostyniuk, who provided technical and editorial feedback; and Judy Settles, Amanda Dallaire, and Lisa Moran, who provided administrative support. Mr. Jim Langford of the Monash University Accident Research Centre provided valuable feedback. All photography is by Gregory Kostyniuk. Dr. Eby thanks Roxanne, Pat, Hal, Beth, and Dan for their love and support. Ms. Molnar thanks Lawrence, Kallen, and Jordan for their encouragement and support, and Liz, Judi, and Esther for their enduring friendship. Ms. Kartje is very thankful to Kirk, Lauren, and Ryan, the aspiring writer in the family, and to her colleagues for their collective inspiration, words of wisdom, and support during the writing of this book.

Acknowledgements

Authors

David W. Eby, PhD is research associate professor and head of the Social and Behavioral Analysis Division of the University of Michigan Transportation Research Institute (UMTRI). Dr. Eby earned a doctorate degree in experimental psychology from the University of California, Santa Barbara, in 1991. An important component of Dr. Eby's research has been to improve the safety and mobility of older adults. He has investigated how in-vehicle navigation assistance systems might be useful for maintaining safe driving in the older population and he led the development and testing of a self-screening instrument intended to educate older adults about their current abilities, what those abilities mean for safe driving, and what older adults can do to continue driving safely. Dr. Eby has directed numerous projects related to transportation and aging. He has received sponsorship for his older adult community mobility research from a variety of sponsors, including the National Highway Traffic Safety Administration, Centers for Disease Control and Prevention, AARP, the Alzheimer's Association, AAA Foundation for Traffic Safety, and the Michigan Office for Highway Safety Planning. Dr. Eby is a former convener of the Transportation and Aging Interest Group of the Gerontological Society of America. He is also the founding Director of the Michigan Center for Advancing Safe Transportation throughout the Lifespan (M-CASTL), a University Transportation Center sponsored by the U.S. Department of Transportation and the University of Michigan. Dr. Eby has authored more than 200 journal articles, technical reports, and book chapters.

Lisa J. Molnar, MHSA is lead research associate with the Social and Behavioral Analysis Division of UMTRI, where she has worked since 1986. She holds a BA in Sociology from Michigan State University and an MHSA in Public Health Policy and Administration from the University of Michigan School of Public Health. Her primary areas of interest are traffic safety and driver behavior. Ms. Molnar has worked on a number of projects focusing on older driver safety and mobility, including the development and testing of two self-screening instruments intended to educate older drivers about how changes in cognitive, perceptual, and psychomotor abilities can affect driving, and what can be done to maintain safe driving. Ms. Molnar authored the guide *Promising Approaches for Enhancing Elderly Mobility* and recently updated it in a project sponsored by AARP, resulting in the publication of *Promising Approaches for Promoting Lifelong Community Mobility*. The new guide contains information on promising programs and other initiatives in several areas including driver screening and assessment, education and training, vehicle design and advanced technology, roadway design, transitioning from independent driving, transportation coordination, and alternative transportation options. She is a coinvestigator on two projects focusing on fitness to drive in early-stage dementia and is the coconvener of the Transportation and Aging Interest Group of the Gerontological Society of America. Ms. Molnar is the Assistant Director of M-CASTL. She has authored more than 100 journal articles, technical reports, book chapters, and conference papers.

Paula S. Kartje, OT, CDRS is a rehabilitation clinical specialist and occupational therapy manager at the University of Michigan Health System, where she has worked since 1986. She holds a BS in Occupational Therapy and became one of the first certified driver rehabilitation specialists in Michigan in 1998. She has experience in a variety of hospital-based areas including acute care, inpatient and outpatient rehabilitation, day treatment, and management. Ms. Kartje has specifically been involved in driver evaluation/rehabilitation for the past 15 years beginning with developing the Drive-Ability Program at the University of Michigan Health System. Ms. Kartje's clinical experience includes evaluating over 100 patients per year, ranging in age from 14 to 95, through this program. This experience has provided her with a vast knowledge of driving problems related to aging and specific medical diagnoses, as well as highly developed evaluation skills. Her primary area of interest is the older driver. She is coinvestigator with UMTRI on a project to develop a Web-based self-screening instrument. She is an advisory board member for the Area Agency on Aging B1-Senior Driver Awareness Program and she has presented at conferences and to multiple groups on driving-related issues.

1 Introduction

The ability to get from place to place is important for everyone. Indeed, a theme from a recent consensus conference on licensing policy for older drivers was "driving is considered a privilege but mobility is a human right" (Molnar & Eby, 2008a). Mobility enables people to conduct the activities of daily life, stay socially connected with their world, participate in activities that make life enjoyable, and increase their quality of life. In most Western nations, personal mobility is equated with the automobile. However, these societies are aging, which can make the operation of a motor vehicle more difficult and potentially less safe. At the same time, people have mobility needs that will still need to be met if driving is limited or no longer possible.

The purposes of this book are to

- Enable readers to understand the issues related to aging and mobility
- Describe the skills related to safe driving and how they can be affected by aging
- Critically examine the current evidence on how medical conditions and medications affect driving skills
- Provide a comprehensive description of driver screening and assessment practices, issues, and tools
- Provide information to help older adults transition from independent driving to driving cessation
- Explore various means by which aging individuals can maintain safe mobility

POPULATION TRENDS

Most industrialized countries are experiencing a dramatic increase in the population of people age 65 and older. In the United States, the number of people in this age group is projected to grow from about 35 million in 2000 to more than 86 million in 2050 (U.S. Census Bureau, 2004). In terms of the percent of the total population, those age 65 and older will account for about 20.7 percent of the population in 2050, up from about 12.4 percent in 2000. Even larger increases are expected from the oldest-old, that is, those age 85 and older. This age group is expected to grow from about 4.3 million in 2000 (1.5 percent of the U.S. population) to 20.9 million in 2050, when they will account for 5 percent of the population (U.S. Census Bureau, 2004).

The growth in the older adult population is expected to be slightly greater for men than for women. U.S. Census Bureau (2004) projections show that women age 65 and

1

older will increase in number by about 2.3 percent between 2000 and 2050, whereas the number of men in this age group will increase by 2.7 percent. These differences are expected to be even greater for those age 85 and older. The number of the oldest-old women is projected to increase by 4.3 percent (13.1 million in 2050) between 2000 and 2050, while the number of men in this group is expected to increase by 6.2 percent (7.7 million in 2050).

What is driving this increase in the older adult population? One factor is that people are living longer. According to the U.S. Census Bureau (2008), *life expectancy* is projected to increase from 76.0 years in 1993 to 82.6 years in 2050. The unknown effects of increasing obesity in U.S. society, however, may lower this projection. In addition to increased lifespan, much of the growth in the older adult population can be attributed to the baby boomers. The baby boomers are the cohort of Americans born during a period of increased birth rates following World War II, between 1946 and 1964 (U.S. Census Bureau, 2006). The first baby boomer will turn age 65 in 2011. By 2029, all living baby boomers will be between the ages of 66 and 83 years of age.

DRIVING TRENDS

DEPENDENCE ON THE PERSONAL AUTOMOBILE

As the baby boomers age, they will bring with them an increased preference for the automobile to meet their mobility needs. Most baby boomers consider driving to be indispensable to their well-being and independence (Molnar & Eby, 2008b). Indeed, in most industrialized countries, mobility is closely linked to the personal automobile. This linkage partly results from the lack of transportation alternatives (Kostyniuk, Shope, & Molnar, 2000) and partly from the culture within which baby boomers grew up. Furthermore, during the years in which baby boomers were first being licensed to drive (approximately 1961–1981), changes in family composition, the tendency to move out of urban areas (suburbanization), and the increased availability and affordability of automobiles made the personal automobile the *de facto* choice for personal mobility (McGuckin & Srinivasan, 2003).

DRIVER'S LICENSE HOLDING

Driver's license holding is high among the current older adult population and is increasing. As can be seen in Figure 1.1, the percentage of males age 65–69 remained around 95 percent between 1993 and 2006, while the licensure rate for males age 70 and above increased slightly during this time period to over 90 percent. The percentage of women holding licenses in the United States, on the other hand, increased dramatically between 1993 and 2006, with about an 8 percentage point increase in licensure for women age 65–69 and a 22 percentage point increase for women age 70 and older. Thus, the licensure rates for women are approaching those of men, and this trend is expected to continue as the baby boomers age (Burkhardt & McGavock, 1999).

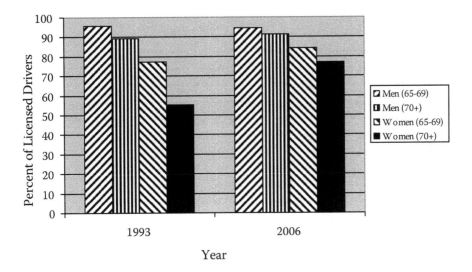

FIGURE 1.1   Licensed drivers as a percent of all people in age group by age group and year (data from Federal Highway Administration, FHWA, 2008a).

CHANGES IN ANNUAL DRIVING DISTANCES

Not only will there be a larger proportion of older drivers holding licenses, they will also be driving more miles. Compared to 1990, the average number of trips per person for people age 65 and older increased from 2.4 to 3.4 in 2001 (Hu & Reuscher, 2004). This increase was greater than for any other age group. The average annual number of vehicle miles traveled for older adults has also increased dramatically (by self-report) for drivers age 65 and older since 1969. Figure 1.2, however, shows that the annual average miles driven per licensed driver apparently reached a plateau for both older men and women between 1995 and 2001 (Hu & Reuscher, 2004). Whether this trend continues as the baby boomers reach older adulthood remains to be seen.

Collectively, the aging population, increases in driver licensure in older adulthood, and the increase in annual miles driven will contribute to the increasing use of the personal automobile for transportation and mobility (Hakamies-Blomqvist, Siren, & Davidse, 2004), even though this reliance is already high. At the same time, those less likely to be driving, such as the oldest women, the poor, those living alone, and racial minorities, will depend greatly on others for rides in an automobile as a passenger (Rosenbloom, 2004).

BEHAVIORAL ADAPTATIONS

It is well established that driving patterns change as people age. These changes result from changes in lifestyle and economic status, and from drivers regulating their driving to compensate for declining abilities (Hakamies-Blomqvist, 2004). Collectively, these changes have been labeled *behavioral adaptations* (Hakamies-Blomqvist,

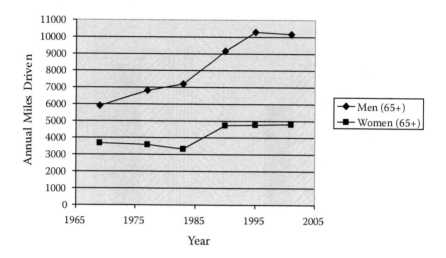

FIGURE 1.2 Average annual self-reported miles driven per licensed driver 65+ years of age by gender and year (data from Hu and Reuscher, 2004).

2004; Smiley, 2004). A large body of literature has shown that when compared to younger drivers, older adult drivers are more likely to avoid difficult driving situations such as nighttime, inclement weather, high traffic times, urban areas, and highways (Ball et al., 1998; Chipman, MacGregor, Smiley, & Lee-Gosselin, 1993; Gallo, Rebok, & Lesikar, 1999; Hakamies-Blomqvist & Wahlström, 1998; Kostyniuk, Shope, & Molnar, 2000; Stamatiadis, Taylor, & McKelvey, 1991). Some older drivers also make adaptations to driving behaviors such as driving slower, driving more often with a passenger, avoiding unprotected left turns across traffic, needing larger traffic gaps for merging, and more frequent use of a safety belt (Ball, Owsley, 1998; Eby, Molnar, & Olk, 2000; Hakamies-Blomqvist & Wahlström, 1998; Keskin, Ota, & Katila, 1989; Van Wolffelaar, Rothengatter, & Brouwer, 1991). However, recent work shows there is considerable variation across studies, making it difficult to determine the extent of self-regulation by older drivers. For example, rates of self-reported avoidance of night driving varies from 8 percent (Baldock, Mathias, Mclean, & Berdt, 2006), to 25 percent (Charlton, et al., 2001), to 60 percent (Ruechel & Mann, 2005), and to 80 percent (Ball, Owsley, et al., 1998). There are also mixed results with regard to the association between self-regulation by older drivers and the functional declines they may be experiencing (Baldock et al., 2006; Ball, Owsley, 1998; Charlton et al., 2001, 2006; Stalvey & Owsley, 2000). While it appears that gender (Charlton et al., 2001; Kostyniuk & Molnar, 2007, 2008; Hakamies-Blomqvist & Wahlström, 1998), awareness and insight into functional impairments (Bal, Owsley, et al., 1998; Freund, Colgrove, Burke, & McLeod, 2005; Owsley, McGwin, Mayes, et al., 2004; Owsley, Stalvey & Phillips, 2003; Stalvey, & Owsley, 2003), and self-perceptions of driving confidence (Baldock et al., 2006; Charlton et al., 2001) are also closely tied to behavioral adaptations, these factors are not consistently examined in studies.

TRAFFIC SAFETY

CRASH RISK

The traffic safety impact of the aging population has received considerable research attention in the past decade (e.g., Transportation Research Board, TRB, 2004). Despite the infrequent, yet highly publicized, fatal crashes involving older adult drivers, there is still some debate as to whether older adults as a group pose a risk to themselves or others on the roadway. In terms of the number of driver deaths, there are far *fewer* fatalities among older adults than younger people. As shown in Figure 1.3., there was a consistent downward trend in fatalities after age 49 in 2006. These data, however, can be misleading because we know that there are fewer people in the older age groups, driver licensure rates drop in the older adult population, and older people drive less than younger people. Each of these factors can lead to fewer total fatal crashes after age 50, independent of declining driving abilities.

To account for these possible effects, it is better to examine motor vehicle fatality data by rates. Figure 1.4 shows driver fatality rates per 100,000 people in each age group; that is, population-based rates. These data show that when corrected for the decreasing number of people in the older age groups, the rate of crashes begins to increase after age 74 and then decrease after age 84. However, as discussed previously, not all people hold licenses, particularly in the youngest and the oldest age groups. As such, population-based rates do not adequately address the risks of fatal crashes by age group. Also plotted in Figure 1.4 are driver fatal crash rates per 100,000 licensed drivers in each age group. In the middle age groups, the rates by licensure are very similar to the population-based rates as nearly all U.S. citizens in the middle age groups hold a license. At the two ends of the age spectrum, however, there are significantly increased fatal crash rates, suggesting that these age groups are at a higher risk of a fatal crash than drivers in the middle age groups.

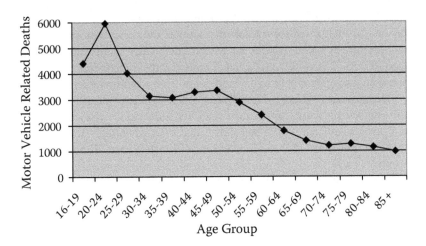

FIGURE 1.3 U.S. motor vehicle driver deaths by age group in 2006 (data from Insurance Institute for Highway Safety, 2007).

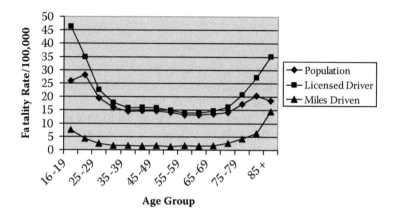

FIGURE 1.4 U.S. motor vehicle driver fatality rates by age group (data from Insurance Institute for Highway Safety, 2007; FHWA 2008a).

Although these data seem to confirm a higher fatal crash risk for older drivers, at the same time we know that older drivers do not drive as much as their younger counterparts (e.g., Smiley, 2004). Indeed, a vast body of literature shows that many older drivers begin to restrict their driving to the times and places that they feel the safest as discussed previously (Ball, Owsley, et al., 1998; Chipman, MacGregor, & Smiley, 1993; Hakamies-Blomqvist & Wahlström, 1998; Stamatiadis et al., 1991). Thus, not only are older drivers reducing their exposure to a crash by driving less, they are also driving in situations where crashes are less likely. To partially account for the reduction in driving for older adults, we calculated fatal motor vehicle crash rates per 100 million miles traveled by age group (2001–2002 data). These data are also plotted in Figure 1.4. When fatal crash rates are considered by miles driven, the rate is higher for drivers age 85 and older than for any other age group.

The data in Figure 1.4 seem to clearly support the conclusion that drivers age 70 and older are at increased risk for a fatal crash when compared to all but the youngest drivers. It is widely recognized, however, that older people are more susceptible than young people to injury and death from a motor vehicle crash, due to frailty (Dejeammes & Ramet, 1996; Evans, 1991; Vivano et al., 1990). The effects of frailty upwardly biases fatal crash rates for older adults. Nevertheless, work by Li, Braver, and Chen (2003) has shown that even when rates are corrected for frailty, older adults are still at a higher risk of crashes than all but the youngest age group.

Recent research from Europe and Australia, however, has questioned this conclusion. This research suggests that the crash rate per mile driven may be biased upward for older drivers because of their tendency to restrict their total miles driven (Hakamies-Blomqvist, Raitanen, & O'Neill, 2002; Hakamies-Blomqvist, 2004; Langford, Fitzharris, Newstead, & Koppel, 2004; Langford, Methorst, & Hakamies-Blomqvist, 2006). This work was based on an original observation by Janke (1991) that drivers who have low annual mileage, regardless of age, have an elevated crash rate when compared to drivers who have high annual mileage. This "low mileage bias" was demonstrated using both Finnish and Dutch data. This research compared self-reported crash involvement by age group and self-reported annual travel distances. This work found that only those older adult drivers who travel less than about 3000 km per year have an elevated crash rate (about 10 percent of the population in the Langford, Methorst, and Hakamies-Blomqvist, 2006, study). Thus, Langford and his colleagues conclude that older adult drivers as a group are not at higher risk of crashes—only those older drivers who drive less than 3000 km annually are high-risk drivers. Recent research in Spain, using self-reported crashes and annual mileage, supported the low mileage bias (Alvarez & Fierro, 2008).

Staplin, Gish, and Joyce (2008) have recommended caution in interpreting this research noting that both the crash and mileage data are self-reported in the studies and participants were self-selected. Staplin et al. present data showing poor reliability of annual mileage estimation within subjects, as well as data showing large errors in estimation of annual mileage when compared to an objective measure (odometer reading). Large underestimation of annual mileage was found for objectively determined low-mileage drivers whereas overestimation was found for objectively determined high-mileage drivers. These systematic errors in self-reported driving distances could explain the low mileage bias. Thus, it is clear that more research, with objectively determined crash and annual data, is needed. Regardless of how this and other issues are weighted, older drivers remain at increased risk for a fatal motor vehicle crash due to a variety of factors including frailty.

This rather complicated picture of older adult traffic safety highlights two important research issues. First, it is very difficult to obtain accurate exposure measures of driving. As just discussed, self-reports are likely to be inaccurate. Even the use of odometer or "black box" data can be flawed if more than one person uses a vehicle. Second, crashes are a relatively infrequent event. To have enough crashes in a sample to make meaningful conclusions, a large number of subjects with driving histories extending over several years are needed. In addition, crash databases generally contain only crashes that are reported to police, missing a large proportion of total crashes. Even for those crashes that are reported, it is difficult to determine who is at fault. A promising approach to overcoming these issues is to instrument a large number of vehicles with technology that determines where and when people are driving (such as a global positioning system, GPS) and the ability to determine who is driving on each trip (such as a face camera). Research along these lines has already started with middle-aged drivers in the "100 car study" (Dingus et al., 1997).

CRASH TYPES

Despite the complexity of these issues, it is clear that older drivers are involved in different types of crashes than younger drivers, as would be expected given the behavioral adaptations common in the older adult population. There is a large body of research documenting that older drivers are over-involved in intersection crashes (Abdel-Aty & Radwan, 2000; Clark, Forsyth, & Wright, 1999; Cooper, 1990; Hakamies-Blomqvist, 1993, 2004; Langford & Koppel, 2006a; Larsen &

Kines, 2002; Li et al., 2003; McGwin & Brown, 1999; Oxley, Fildes, Corben, & Langford, 2006; Staplin, Lococo, McKnight, McKnight, & Odenheimer, 1998; Zhang, Lindsay, Clarke, Robbins, & Mao, 2000). A study in Australia (Langford & Koppel, 2006a) looked at national fatal crash data for the period 1996–1999 and compared crashes among three driver age groups: middle (40–55), young-old (65–74), and old (75 years and older). They found that the percentage of fatal crashes at intersections was 50 percent for the old age group compared to only 21 percent for the middle age group, while the young-old had 35 percent of fatal crashes at intersections. (It is interesting to note that the study also found no differences in fatalities at roundabouts, with only two fatalities across all age groups.) Similar work in the United States using national fatality data has found that when compared to middle-aged drivers, U.S. drivers age 65–69 were 2.3 times more likely to be in an intersection crash, whereas drivers age 85 and older were 10.6 times more likely to be in an intersection crash (Preusser, Williams, Ferguson, Ulmer, & Weinstein, 1998).

Older drivers are also more likely to get into multiple-vehicle crashes when compared to younger drivers (Langford & Koppel, 2006a; Cooper, 1990; Hakamies-Blomqvist, 1993). Langford and Koppel (2006a) found that 74 percent of older driver fatal crashes involved multiple vehicles compared to 60 percent for middle-age drivers. This outcome is not surprising given the high percentage of intersection crashes among older adults and that older drivers are underrepresented in alcohol- and illicit drug-related crashes (Eby, 1995; Langford & Koppel, 2006a; Hakamies-Blomqvist, 1993). Older drivers are also less likely to engage in risk-taking behaviors that often lead to single-vehicle crashes such as those caused by speeding.

The timing and location of older driver crashes is also different than for younger drivers. Older driver fatal crashes occur more often during the daytime, during off-peak hours, on dry roads, and on nonexpressway/low-speed roads (Langford & Koppel, 2006a; Cooper, 1990; Hakamies-Blomqvist, 1993). These findings are in agreement with what we know about older driver behavioral adaptations—older drivers tend to travel during these times and at these locations.

Perhaps the most important difference between older and younger drivers is that older drivers are more likely to be at fault in crashes. A study of more than 14,000 crashes in Canada found that when compared to middle-age drivers (36–50 years of age), young-old (65–74) and old drivers (75 and older) were significantly more likely to be responsible for causing the crash (Cooper, 1990). A study of more than 1,000 fatal older driver crashes in Finland found that drivers age 65 and older were responsible for 87 percent of their multiple-vehicle crashes whereas middle-aged drivers were responsible for only 50 percent of their multiple-vehicle crashes (Hakamies-Blomqvist, 1993).

MOBILITY NEEDS

Given older drivers' increased crash risk and tendency to be responsible for crashes, what should society do? The issue of older driver safety is surrounded by a fair amount of confusion, emotions, and naïve solutions. One solution often voiced is to simply "get old folks off of the road" (see e.g., Carr, 2000). This solution ignores the facts that only a portion of older drivers are unsafe, some older drivers can improve their skills through training, and older people, like all people, have mobility needs that still need to be satisfied if driving is no longer allowed.

Like all drivers, older drivers are reluctant to give up driving and consider it to be essential to independence and quality of life (Carp, 1988; Kaplan, 1995). Driving provides an opportunity for older adults to stay engaged in their community and to participate in activities that enhance their well-being, particularly in areas where alternative transportation options are limited. A number of recent studies suggest that driving cessation is associated with increased depressive symptoms over time and declines in general psychological well-being (e.g., Fonda, Wallace, & Herzog, 2001; Marottoli et al., 1997; Ragland, Satariano, & MacLeod, 2005). Given the reliance on and preference for personal automobile travel, one's license should be taken away only after a comprehensive assessment shows that a person can no longer drive safely and that he or she cannot benefit from rehabilitation. At the same time, society must recognize that alternatives to the personal automobile for transportation are generally poor in most areas and considered unacceptable to older adults (see, e.g., Kostyniuk, Shope, & Molnar, 2000). The mobility needs for older adults who can no longer drive still need to be met.

As such, two complementary but interdependent goals have emerged with respect to older drivers: to help those who are able to drive safely continue to do so, and to identify and provide community mobility support to those who are no longer able to drive (Molnar, Eby, & Dobbs, 2005). These goals essentially represent two ends of an older adult transportation and mobility continuum as described by Molnar and Eby (2008c). At one end of the continuum is safe independent driving, with research and program efforts focusing on crash and injury prevention (see Figure 1.5). At the other end is dependence on nondriving transportation options, with research and program efforts focusing on the maintenance of older adult mobility. Bridging these two ends is the transitioning process from driving to nondriving, which encompasses both safety and mobility issues and cuts across many points along the continuum.

Also shown in Figure 1.5 is how each chapter in this book fits into the older adult mobility framework. Some of the chapters span the whole continuum, while others are directly related to only one part of the continuum. The book starts with a discussion on how aging can affect functional abilities and conceptualization of the skills needed for safe driving. Next we discuss how medical conditions and medications can cause functional declines that can affect safe driving. The next several chapters discuss the issues surrounding the screening, assessment, and licensing of older drivers. The book continues with discussions of several issues related to maintaining safe driving given age-related declines, including education/rehabilitation, vehicle adaptations and advanced technology, and roadway design. Finally, we conclude with discussions of how to maintain safe mobility when driving is no longer desirable or possible, including driving retirement and the transition to dependence on other community mobility options.

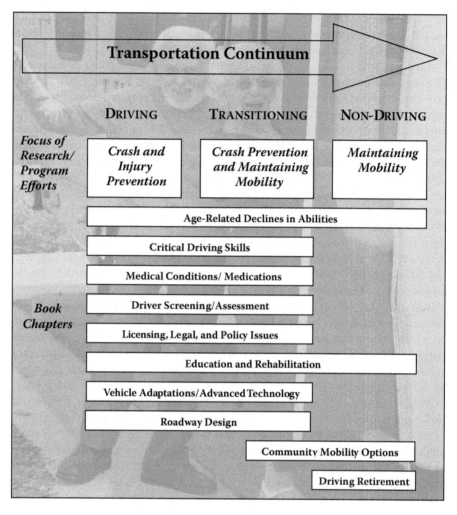

FIGURE 1.5   The continuum of transportation, the changing focus of research and programs along the continuum, and the parts of the continuum addressed by the book chapters.

2 Age-Related Declines in Abilities

There is still no cure for the common birthday.

John Glenn

INTRODUCTION

Driving is a complex task that requires visual, cognitive, and psychomotor abilities. As people age, they may experience declines in these functional abilities as a result of medical conditions that become more prevalent with age, the medications used to treat them, and the aging process itself that can compromise their ability to drive safely. While individual differences are large (Department of Transport, 2001; European Road Safety Observatory, 2006; Janke, 1994) and the impacts of such declines on actual crash risk are not always fully known (Whelan, Langford, Oxley, Koppel, & Charlton, 2006), it is clear that with increasing age most older adults experience some loss in visual perception ability (e.g., Bailey & Sheedy, 1988; Owsley & Sloane, 1990; Schieber, 1994), decreases in cognitive functioning (e.g., Cerella, 1985; Denney & Palmer, 1981), and/or decreased psychomotor function (e.g., Kausler, 1991; Marottoli & Drickamer, 1993; Yee, 1985). Thus, it is important to understand the key abilities related to psychomotor functioning, vision, and cognition, and how they are affected by the aging process.

This chapter reviews findings from the literature relative to how psychomotor, visual, and cognitive abilities decline with aging, and the implications of these declines for driving performance and, if known, actual crash risk. This review builds on and extends earlier work by Eby, Trombley, Molnar, and Shope (1998). While it examines each psychomotor, visual, and cognitive ability separately, it is important to recognize that a decline in one area may interact with or exacerbate the effects of a decline in another area, although declines in one area do not necessarily predict declines in another (Department of Transport, 2001). Thus, the three areas of abilities act as a system that influences driving safety to a greater or lesser extent depending on a whole host of environmental, vehicle, and driver characteristics explored throughout this book (Fjerdingen et al., 2004).

PSYCHOMOTOR ABILITIES

Psychomotor functioning refers to an individual's coordinated and controlled ability to move and orient parts of his or her body (Kelso, 1982). Declines in psychomotor abilities can make it difficult for drivers to get in and out of, or operate, a motor vehicle and can influence injury and recovery in the event of a motor vehicle crash (Sivak et al., 1995). Psychomotor abilities tend to decline with increasing age. This section reviews the effects of age on the speed at which movements are initiated and completed (reaction time), the range of motion that is possible (flexibility), the accuracy of movements (coordination), and the forces required to execute the movement (strength).

REACTION TIME

One widely recognized age-related change is increased reaction time among older adults, resulting from a slowing of central-processing functions and joint stiffness, which decreases muscle power and simple reflex movements (Department of Transport, 2001; Klavora & Heslegrave, 2002). Increased reaction time can lead to difficulties in driving, especially in unfamiliar or congested areas (Suen & Mitchell, 1998). Two types of reaction time have implications for safe driving: simple and choice. Simple reaction time involves an individual making one response to a single stimulus. Choice reaction time involves an individual distinguishing among two or more stimuli and possibly having one or more responses to make (Marottoli & Drickamer, 1993). Both types of reaction times clearly increase with age. There is also a greater difference between young and older adults for choice reaction time than for simple reaction time (Marottoli & Drickamer, 1993); that is, as task demands increase, older adults take increasingly longer to respond than young adults. The age difference in reaction time is particularly relevant for the complicated tasks and decisions that must be made quickly while driving. Mihal and Barrett (1976) found that as choice reaction time increased, so did motor vehicle crash involvement. The same relationship was not found for simple reaction time. Ranney and Pulling (1989) found that increased choice reaction time among drivers age 75 and older had a strong association with overall driving performance and measures related to vehicle control. However, correlations between measures of reaction time and crash history were weak.

FLEXIBILITY

Joints and muscles have a physiologically determined range through which they can move. As individuals age, physiological changes occur in the musculoskeletal system that can affect driving ability. Older adults with less joint flexibility have been found to exhibit poorer on-road driving ability than those with wider ranges of motion (McPherson, Michael, Ostrow, & Shafron, 1988). Joint flexibility changes as a precursor to various forms of arthritis and other health conditions. Muscle strength also decreases by age 55, and musculature may be tighter because of decreases in active stretching from heavy manual labor, sports, or stretching exercises that occur with age (States, 1985). Loss of flexibility in the limbs may affect a driver's ability to quickly shift his or her right foot from the accelerator to the brake or to safely maneuver the vehicle through turns and around obstacles (Staplin, Lococo, Stewart, & Decina, 1999).

A common age-related decline in flexibility has to do with head rotation (Malfetti, 1985). In many driving situations, drivers must be able to direct or redirect their gaze in many different directions to check for potential conflicts (e.g., to make the familiar "left-right-left" check before crossing an intersection or look over their shoulder before merging with traffic or changing lanes; Staplin et al., 1999). Restrictions in range of neck motion can impede drivers' ability to scan to the rear, back up, and turn the head to observe blind spots (Janke, 1994; Malfetti, 1985). Reverse parking can be difficult when drivers are unable to look over their shoulder (Bulstrode, 1987). In a survey of 446 older drivers, 21 percent reported that it was "somewhat difficult" to turn their heads and look to the rear when driving or backing (Yee, 1985). Most drivers in a study by Bulstrode (1987) reported that the original interior rearview and near-side exterior door mirrors did not compensate for their limited neck mobility. There is evidence that declines in head rotation ability increase crash risk, with one study showing a twofold increase in crash risk (Marottoli et al., 1998).

COORDINATION

Psychomotor behavior involves not only adequate flexibility and reaction time but also precision of movement or coordination. Coordination of arm and leg motion is important to control the vehicle (Wheatley, Pellerito, & Redpenning, 2006). Older adults have less accuracy in movement than younger adults (Anshel, 1978; Marshall, Elias, & Wright, 1985; Szafran, 1953; Welford, 1959). Szafran (1953) had young and older individuals move their hand sideways for a set distance and found that the magnitude of hand movement error was approximately one-third greater for those subjects age 50 to 69 than for younger individuals. However, a discrete task like that in the hand movement study may not be related to continuous tasks like driving, in which movement actions are continuously altered based on sensory feedback (Kausler, 1991). Studies have shown that the accuracy of continuous movements showed an even greater decline with age than did discrete movements (Ruch, 1934; Snoddy, 1926; Wright & Paine, 1985). For example, Ruch (1934) had young and older adult subjects try to keep a stylus directly pointed at a spot on a disk; when the stylus pointed at the spot, the disk rotated. Thus the task was to keep the disk rotating, and the precision measure was the number of disk rotations in 30 seconds. Ruch's results showed that the mean number of rotations for individuals age 60–82 was 82 percent of the mean for individuals in younger age groups. Thus, it is clear that the precision of discrete and continuous movements decreases after about age 60. The influence of these age-related deficits on driving performance and crash involvement has not been established.

STRENGTH

Physical strength is also considered important for safely operating a vehicle, particularly hand, shoulder, and leg strength. For example, loss of strength in the lower limbs may affect an individual's ability to apply correct pressure for appropriate speed control (Staplin et al., 1999). One clear effect of aging is a decrease in muscle strength. After about age 40, muscle strength begins to decline, with a

25 percent decrease by age 65 (Petrofsky & Lind, 1975; Shepard, 1998). Loss of strength continues with increasing age and eventually impedes activities of daily living. The relationship between crashes and decreased muscle strength has not received adequate research attention. However, studies using surrogate measures of strength, such as walking speed, have found that those drivers who walk more slowly have more safety-related errors while driving (Eby & Molnar, 2001) and are more likely to be involved in adverse traffic events (Marottoli, Cooney, Wagner, Doucette, & Tinetti, 1994; Staplin, Gish, & Wagner, 2003). Muscle strength has been shown to increase in a short period through a proper weight-training regimen (Fiatarone et al., 1990).

Anstey, Wood, Lord, and Walker (2005) reviewed several studies that examined the effects of physical function on actual driving. Apart from declines in neck rotation ability (described earlier), findings relative to other physical functions were not conclusive with regard to crash risk. The authors surmised that this may be due to a lack of studies examining physical function on crash risk, poor measurement of disease severity, and the possibility that older individuals with physical limitations such as reduced strength or flexibility may be more likely to be aware of their impairment than older adults with visual or cognitive impairment and stop driving on their own.

VISUAL ABILITIES

Driving is a highly visual task, with most of the information that drivers process being visual (Klavora & Heslegrave, 2002). Declines in visual abilities become more common with increasing age (Attebo, Mitchell, & Smith, 1996) through both the normal aging process and the increased prevalence of eye disease (Anstey, Wood, Lord, & Walker, 2005). This section describes several visual abilities that can affect driving, including anatomic changes, eye movements, sensitivity to light, dark adaptation, visual acuity, spatial contrast sensitivity, visual field, space perception, and motion perception.

ANATOMIC CHANGES

As individuals age, they may experience anatomic changes in their eyes that can lead to greater difficulty seeing while driving. For example, the amount of light reaching the retina (retinal illuminance) appears to be markedly decreased in older adults (Corso, 1981; Owsley & Ball, 1993; Owsley & Sloane, 1990; Schieber, 1994). This may occur for two reasons. First the maximum diameter of the pupil decreases with increasing age (Lowenfeld, 1979; Weale, 1971); the smaller the pupil, the less available light that can enter the eye. In dim illumination conditions, the average pupil diameter of 20-year-olds has been shown to be about 7 mm compared with only 4 mm for 80-year-olds (Lowenfeld, 1979). A second reason for decreased retinal illuminance in older adults is an increase in the absorption of light that enters the eye (Boettner & Wolter, 1962; Said & Weale, 1959; Sarks, 1975; Stocker & Moore, 1975). When light enters the eye, it passes through several relatively transparent structures and media, including the cornea, aqueous humor, crystalline lens, and vitreous humor. An increase in opacity in any of these ocular structures increases the amount of light that is absorbed, thereby decreasing the amount of light reaching the light-sensitive retina. With increasing age, a corneal grayish-yellow ring begins to develop in about 75 percent of the population (Stocker & Moore, 1975); corneal opacity slightly increases (Boettner & Wolter, 1962; Stocker & Moore, 1975); crystalline lens opacity increases (Said & Weale, 1959; Sarks, 1975; Coren & Girgus, 1972; Spector, 1982); and the formation of condensations in the vitreous, known as *floaters*, increases (Corso, 1981). Collectively, these changes reduce the amount of light reaching the retina of older adults to approximately one-third the light reaching the retina of a 21-year-old (Weale, 1982). Thus, during nighttime driving, an older driver requires brighter lights than a young driver to see well.

When light enters the eye, it bends as it passes through the cornea and crystalline lens (Westheimer, 1986). The amount of bending can be controlled by changing the shape of the crystalline lens to focus an image on the retina. Objects that are closer require greater focusing to be clearly seen than objects that are far away. This process, called *accommodation*, is disrupted by age-related declines in the ability to change the shape of the lens in a condition known as *presbyopia* (Corso, 1981). Studies suggest that by about age 65, the crystalline lens has lost most of its ability to accommodate (Hofstetter, 1965). Thus, older drivers have difficulty clearly seeing objects that are nearby, such as a vehicle dashboard. Individuals with presbyopia are generally prescribed corrective lenses (bifocals or trifocals) that help them accommodate to near distances (Owsley & Ball, 1993). However, the near correction is usually set for reading distance (approximately 40 cm), whereas motor vehicle controls are typically greater than 40 cm away, leading to potential difficulty in reading dashboard displays.

Eye Movements

The ability to resolve fine spatial detail is not uniform across the retina. A small region in the retinal center, known as the *fovea*, is densely packed with special cells (cone photoreceptors) that have the greatest ability to resolve fine spatial detail (Matlin & Foley, 1992). When an individual looks directly at an object, the image of the object falls on the fovea. As such, the ability to see fine detail is partially dependent on eye movement ability to keep an object's image centered on the fovea. There is evidence of age-related declines in eye movement ability. Studies have found that older adults have an increased latency for saccadic eye movement (short-duration and high-velocity movements); that is, it takes older adults longer to start a saccadic movement than it takes younger adults (Abel, Troost, & Dell'Osso, 1983; Spooner, Sakala, & Baloh, 1980; Wacker, Busser, & Lachenmayr, 1983; Warabi, Kase, & Kato, 1984). It also appears that saccadic movement velocity is slower in older adults than in younger adults (Abel et al., 1983; Spooner et al., 1980; Warabi et al., 1984). In addition, older adults typically require more saccadic eye movements than younger adults to fixate an image on the fovea (Lapidot, 1987). Collectively these results show that it takes older adults longer to locate objects in the visual scene, although the accuracy of saccadic eye movements does not seem to be affected by age (Warabi et al., 1984). However, pursuit eye movements (long-duration and slow-velocity movements) do show age-related declines. When compared with young adults, older adults show significantly slower pursuit velocities and decreased latencies for onset of pursuit movements (Lapidot, 1987; Sharpe & Sylvester, 1978). Sharpe and Sylvester (1978) discovered that young people could accurately track targets moving at velocities up to 30 deg/sec, while older adults could accurately track targets only up to a velocity of 10 deg/sec. This implies that older drivers have more difficulty than younger drivers resolving the details of objects that are in motion.

In addition to its effects on saccadic and pursuit eye movements, aging also restricts the maximum extent of gaze without head movement (Chamberlain, 1970, 1971; Huaman & Sharpe, 1993). Chamberlain found that the maximum extent of upward gaze for the 75–84 age group was less than half the maximum extent found in the 5–14 age group. Similar results have been found for downward gaze extent (Huaman & Sharpe, 1993). There do not appear to be any age differences for left- or right-eye movement extents. These results show that older drivers may have to initiate head movements to read the dashboard after looking at the road, while younger drivers only need to move their eyes.

Visual Acuity

Visual acuity refers to the ability to perceive spatial detail, such as a road sign, at a given distance (Olzak & Thomas, 1986). Good visual acuity is important for the perception of traffic signs and signals, as well as for long-distance viewing needed for driving maneuvers such as passing on secondary roads (European Road Safety Observatory, 2006). It is clear that static visual acuity (when objects and the driver are not moving) begins to decline from normal levels at about age 40 to 50 (Owsley & Sloane, 1990) and continues to decline through at least age 90. Based on the results of several studies, Verriest (1971) concluded that acuity decreases from about 20/20 for 50-year-olds (acuity is better than normal in the younger age groups) to about 20/60 in 90-year-olds. Thus as individuals age, spatial details, such as letters, have to be increased in size to be seen at the same distance as when they were younger. For individuals of all ages, static acuity improves when the overall stimulus illumination is increased and the contrast between the stimulus and background is increased (Corso, 1981). A related type of acuity is dynamic visual acuity; that is, the ability to resolve fine detail when there is relative motion between the stimulus and the observer, as in a driver in a moving vehicle reading a traffic sign (Burg, 1966). Dynamic visual acuity has been shown to decline with age in much the same way as static visual acuity (Burg, 1966; Burg & Hurlbert, 1961; Heron & Chown, 1967; Long & Crambert, 1989). However, the decline tends to start at an earlier age and tends to be steeper than the decline in static visual acuity. Burg (1966) found that when compared with 20-year-olds, dynamic visual acuity of 70-year-olds had declined by about 60 percent. Declines in visual acuity often occur so slowly that individuals may not notice their worsening ability, especially in poor light (European Road Safety Observatory, 2006). This can lead to an overestimate of visual powers and failure to compensate for declines in visual acuity (e.g., by appropriately wearing prescription glasses; Department of Transport, 2001).

Contrast Sensitivity

Contrast sensitivity has to do with the ability to detect a target against its background by means of slight differences in luminance or reflectance (Smiley, 2006). Specifically, it refers to the amount of difference between light and dark parts of a pattern of a certain size required for an individual to detect the pattern (Owsley & Sloane, 1990). The size of the pattern is often defined by spatial frequency and contrast sensitivity and is typically studied using gratings that vary sinusoidally in luminance. Low-frequency gratings have few changes in luminance contrast (cycles) for a given size, whereas high-frequency gratings have many. For a given cycle the smallest difference between the maximum and minimum luminance that people can reliably detect can be directly converted to their contrast sensitivity for that spatial pattern. Contrast sensitivity allows an individual to see targets that do not differ greatly in brightness or color from the surrounding background area or that may have "fuzzy" or ill-defined edges (e.g., when the edge of the road has a worn/faded or missing edgestrip or when the color of the shoulder is similar to that of the paved surface; Staplin et al., 1999).

Contrast sensitivity is affected by aging. Older adults with healthy, normal eyes show a marked decline in contrast sensitivity for high-frequency gratings (Derefeldt, Lennerstrand, & Lundh, 1979; Kline, Schieber, Abusamra, & Coyne, 1983; Madden & Greene, 1987; Owsley, Sekuler, & Siemsen, 1983; Schieber, Kline, Kline, & Fozard, 1992) with notable declines for individuals over age 60 starting at frequencies of 2 cycles/deg visual angle.

Sensitivity to Light

In a darkened environment, such as in a car at night, dim lights may be difficult to see. The dimmest light that individuals can see some percentage of time (usually one-half of the time) is referred to as their *sensitivity to light* (Hood & Finkelstein, 1986). There is good evidence that visual sensitivity decreases dramatically with age. Sensitivity to light is typically studied after being in a darkened environment for approximately 30 minutes. In these conditions, it has been shown that sensitivity decreases with age; that is, the older the individual, the brighter the light must be to be seen (Birren & Shock, 1950; Domey, McFarland, & Chadwick, 1960; McFarland & Fischer, 1955; McFarland, Domey, Warren, & Ward, 1960). In one study (McFarland et al., 1960), the sensitivity of 20-year-olds was 200 times greater than the sensitivity of 80-year-olds.

Dark Adaptation

When individuals move from a bright to a dark environment, such as walking into a darkened movie theater or driving into a mountain tunnel, they may have difficulty seeing dim lights at first. Then after a few minutes, the dim lights become easier to see. This process of increasing sensitivity to light with increasing length of time in the dark is known as *dark adaptation* (Hood & Finkelstein, 1986). Findings are inconclusive with respect to the rate at which the eyes adapt to the dark, with some studies showing that the rate is slower for older than for younger individuals (Birren & Shock, 1950; Domey et al., 1960; Eisner, Fleming, Klein, & Mauldin, 1987; McFarland, 1968; McFarland et al., 1960) and others finding no age differences (Birren & Shock, 1950; Eisner et al., 1987). However, glare recovery time, a related visual function, has been shown to increase with age (Brancato, 1969). Glare occurs when light enters the eye in such a way as to temporarily disrupt vision (Corso, 1981). In nighttime driving glare occurs, for example, when a passing car's headlights shine into a driver's eyes or headlights are viewed in a driver's rearview mirror; during the day sun glare can cause similar problems (Staplin et al., 1999). Brancato (1969) found that the time required for glare recovery was approximately 9 seconds for 65-year-olds compared with about 2 seconds for 15-year-olds. Other work has shown that the debilitating effects of glare are greater for older drivers (Wolf, 1960). In a test of ability to see an object, Wolf (1960) found that after a glare stimulus, older adults required the object to be significantly brighter than did younger people to be seen. The 75–80 age group needed the object to be 50 to 70 times brighter to be reliably seen than did the 5–15 age group. Thus, drivers age 65 and older would have greater difficulty seeing after having headlights flashed in their eyes and would take longer to recover from the glare than younger drivers.

Visual Field

Visual field refers to the "extent of visual space over which vision is possible with the eyes held in a fixed position" (Sekuler & Blake, 1985, p. 499). The larger the visual field, the more individuals can see without moving their eyes. Vision performance in the periphery of the visual field (peripheral vision) is poorer for older adults than for young adults (Burg, 1968; Crassini, Brown, & Bowman, 1988; Wolf, 1967). Studies have shown that shrinkage of the visual field with age is much more pronounced when the person performs a distracting task or distracting stimuli are presented with the target (Ball, Beard, Roenker, Miller, & Griggs, 1988; Scialfa, Kline, & Lyman, 1987; Sekuler & Ball, 1986; Staplin et al., 1999; Triesman & Gelade, 1980). Thus when attentional demands are placed on an older person, the size of the visual field that can be used (the useful field of view [UFOV®]) is reduced. Ball et al. (1988) found that the UFOV® in an older adult can be reduced to one-third that of a young adult.

Space Perception

The ability to perceive the relative distances of objects and the absolute distance from an object to the observer is known as *space perception*. The ability to accurately perceive space allows drivers to know how much distance is between their vehicle and the vehicle ahead of them or the amount of space in a traffic gap for merging with or crossing a traffic stream. Correct judgments of gaps in traffic for turning and merging depend strongly upon how quickly and accurately an individual's brain can interpret changes in the size of the image that is formed on the retina at the back of the eye when his or her gaze is focused on a distant object (Staplin et al., 1999).

Given the frequency with which older drivers are involved in intersection crashes (Vivano, Culver, Evans, Frick, & Scott, 1989) and left-turn crashes (NHTSA, 1993), one may suspect that older drivers have deficient perception of space. Unfortunately few studies have investigated the effect of aging on space perception. Only one type of space perception, stereopsis, has received much research attention. Stereopsis is a source of depth information that uses the retinal images from both eyes to determine depth and distance relationships (Arditi, 1986; Matlin & Foley, 1992). The general finding in the literature indicates that stereopsis, as measured by either a stereoscope or a random-dot stereogram, declines significantly with age in people over age 40 (Bell, Wolf, & Bernholz, 1972; Hoffman, Price, Garrett, & Rothstein, 1959; Hofstetter & Bertsch, 1976; Jani, 1966). However, because of methodological and reporting problems, these studies are not conclusive (Owsley & Sloane, 1990). Thus it is not known how space perception is affected by aging.

MOTION PERCEPTION

Objects in the environment, such as vehicles, people, birds, and trees, frequently move from one location to another or change their shape. The ability to perceive these changes is known as *motion perception*. Because driving creates and takes place in a dynamic environment, adequate motion perception is important for safe and efficient driving. Studies have shown that certain kinds of motion perception may decline with age and that sensitivity to motion (i.e., the ability to detect small motions) may be lower in older than in young adults, with older adults requiring more motion than young people to see the motion (Ball & Sekuler, 1986; Schieber, Hiris, White, Williams, & Brannan, 1990; Trick & Silverman, 1991). However, other work has suggested no age-related decline or that the decline in motion sensitivity may be restricted to older women (Schieber et al., 1990; Brown & Bowman, 1987; Gilmore, Wenk, Naylor, & Stuve, 1992). Older drivers may also have greater difficulty perceiving motion in depth, an important process involved in drivers' knowing the change in position of their vehicle relative to the vehicle in front of them. Hills (1975) found that the ability to detect the movement of two lights moving away from each other (as would happen perceptually with two taillights as a driver approached) declined after about age 60, especially during simulated night driving conditions. Although Hills (1975) interpreted these findings as a decline in sensitivity to angular displacement, subjects most likely perceived the stimuli as moving in depth rather than moving apart. Other studies have shown that the ability to detect and accurately perceive the motion of an object in depth declines with age (Shinar, 1977; Schiff, Oldak, & Shah, 1992). Older drivers also have greater difficulty than young drivers detecting the relative speeds of objects but are more accurate at judging the absolute speed of a vehicle (Hills, 1980; Scialfa, Guzy, Leibowitz, Garvey, & Tyrrell, 1991). Collectively these results suggest that older drivers may not perceive critical, motion-defined traffic situations as quickly as younger drivers and therefore have less time to react when these situations arise.

THREE-DIMENSIONAL STRUCTURE FROM MOTION

Object shape and depth can be perceived solely from motion information in a phenomenon known as *three-dimensional structure from motion* (3-D SFM) or the kinetic depth effect (Wallach & O'Connell, 1953). SFM is studied by simulating animated 3-D objects constructed of points of light on a computer screen. Thresholds for detecting SFM are usually determined by adding points of light that are moving independent of the simulated object points ("noise") and determining the signal-to-noise level where the object can be accurately identified some set percent of the time (e.g., Braunstein, 1986). Several studies have found that when compared to younger people, older adults had higher thresholds for detecting SFM, were less accurate at identifying 3-D SFM objects, and had poorer sensitivity (Andersen & Atchley, 1995; Norman, Dawson, & Butler, 2000; Norman & Ross, 2003).

COGNITIVE ABILITIES

As part of the normal aging process, there are declines in many cognitive abilities that are needed to perform complex tasks such as driving (Anstey et al., 2005). It is individuals' cognitive processes that allow them to take visual cues in the environment and select the appropriate information, interpret it, and make decisions that must then be translated into appropriate driving actions (European Road Safety Observatory, 2006). The key cognitive abilities that can affect driving include attention, memory, problem solving, and spatial cognition.

ATTENTION

Attention refers to the process of concentrating a limited cognitive resource to facilitate perception or mental activity (Anderson, 1985; Bernstein, Roy, Srull, & Wickens, 1991; Matlin, 1989). Although many cognitive abilities are needed for driving and no single ability will be predictive of all aspects of driving, evidence suggests that inattention is strongly related to safe driving (Department of Transport, 2001). Driver inattention has been found to be a major factor in traffic crashes, with 20–50 percent of crashes involving some form of inattention (Goodman, Bents, Tijerina, Wierwille, Lerner, & Benel, 1997; Ranney, Garrott, & Goodman, 2001; Stutts, Reinfurt, & Rodgman, 2001; Sussman, Bishop, Madnick, & Walter, 1985; Wang, Knipling, & Goodman, 1996). One form of inattention is driver distraction, which results from a triggering event (Stutts et al., 2001). A distracted driver has delayed recognition of information necessary for safe driving because an event inside or outside of the vehicle has attracted the driver's attention (Stutts, Reinfurt, & Rodgman, 2001). A distracted driver may be less able to respond appropriately to changing road and traffic conditions, leading to an increased likelihood of crash. Driver distraction has been estimated to be a contributing factor in 8 to 13 percent of tow-away crashes (Stutts et al., 2001; Wang et al., 1996).

Two types of attention are reviewed here: divided and selective. Divided attention occurs when a person monitors two or more stimulus sources simultaneously or performs two tasks simultaneously (Parasuraman, 1991). Driving situations in which divided attention is required are numerous. Attentional abilities come into play when a driver must monitor and respond effectively to multiple sources of information at the same time (e.g., a driver entering a freeway must track the curvature of the ramp and steer appropriately, keep a safe distance behind the car ahead, and check for gaps in traffic on the highway, while simultaneously accelerating just enough to permit a smooth entry into the traffic stream; Staplin et al., 1999). Distractions inside and outside the vehicle may compound the problem.

Crash statistics and observational studies suggest that older drivers are particularly hindered by situations that require divided attention (e.g., turning left at an intersection and perceiving relevant traffic signs; Mihal & Barrett, 1976; Kahneman, 1973; van Wolffelaar, Brouwer, & Rothengatter, 1991). Although divided attention ability is poor for individuals of all ages (Ponds, Brouwer, & van Wolffelaar, 1988), older adults show a significantly decreased ability to divide attention when compared with young and middle-age adults (Ponds et al., 1988; Salthouse, Mitchell, Skovronek, & Babcock, 1989). These age differences, found in laboratory experiments, may be more pronounced in actual traffic situations. Greater problems with divided attention for older drivers may be expected in real-life traffic situations because the laboratory tests do not include an active visual search for information at unpredictable locations; this type of visual search is especially age sensitive (Ponds et al., 1988). To assess divided attention ability in actual traffic conditions, Crook, West, and Larrabee (1993) developed a test requiring older drivers to drive and to monitor weather and traffic reports played on a radio. Significant declines were found for the oldest age groups on driving performance and recall of weather and traffic reports when drivers had to pay attention to both tasks at the same time. These results suggest that the ability to simultaneously attend to more than one stimulus or task is poorer for older drivers than for drivers in younger age groups.

Selective attention is the ability to ignore irrelevant stimuli while focusing attention on relevant stimuli or tasks (Parasuraman, 1986, 1991). To drive effectively, individuals must be able to ignore the hundreds of sensations impinging on their perceptual systems and focus their attention on the control of the vehicle, dashboard information, and the movement of nearby vehicles. They must also be able to quickly shift their attention among several important stimuli, a task called *attention switching* (Parasuraman, 1986). Although drivers are not conscious of it, they must constantly filter out millions of times as much sensory information as they actually use to make decisions and control their vehicles; thus, intact selective attention abilities are essential to anticipate and respond appropriately to hazards (Staplin et al., 1999).

MEMORY

Memory is the mental process whereby people store their knowledge and experiences. Because it is the process that allows drivers to recall traffic laws and driving skills, predict traffic situations, and determine their location, good memory ability is essential to safe and efficient driving. Drivers must think about and recall recently learned information while driving, without any lapses in safely controlling their vehicle, such as remembering and applying a simple set of navigational instructions memorized before a journey while driving in heavy traffic (Staplin et al., 1999).

Older adults report more problems with their memory than younger individuals (Cutler & Grams, 1988; Ryan, 1992). Studies controlling for declining health have found that people over age 60 perform more poorly on memory tasks than younger people (Bahrick, 1984; Hultsch, Hammer, & Small, 1993; Maylor, 1991; Perlmutter & Nyquist, 1990; Rabbitt, 1989; West, Crook, & Barron, 1992). The reason for the decline in memory performance is not well understood (Light, 1996). Two types of memory processes are particularly important in driving: working memory and long-term memory.

Working memory (WM) is used to conduct ongoing cognitive activities (Baddeley, 1984). WM is the conscious part of memory where thinking takes place (Siegler, 1991) and is therefore critical for driving. The capacity of WM is limited (Miller, 1956); that is, only a certain amount of information can be considered simultaneously. Numerous studies, using several different WM tasks, have shown that older adult performance is worse than that of young adults (Salthouse et al., 1989; Salthouse, 1987, 1990; Salthouse & Skovronek, 1992). These results suggest that some decline in memory ability for older people comes from a reduction in the capacity of WM or in the ability to effectively organize information (Kausler, 1991; Taub, 1974).

Information in WM also has a limited duration (Brown, 1958). If individuals are prevented from rehearsing the information in WM, that information will be forgotten. Typically the duration of WM is determined using the Brown-Peterson task; that is, by showing individuals a stimulus that does not exceed WM capacity, like the letters *HKP*, and then asking them to recall the letters after waiting for a variable period of time (Peterson & Peterson, 1959). Rehearsal is prevented by having individuals perform some other mental task, like counting backward, during the waiting period. WM duration is determined by how accurately individuals can recall information after the various waiting periods. Schonfield (1969) found that older adults' performance on the Brown-Peterson task was poorer than the performance of young adults, with significant forgetting occurring after only 6 seconds.

There also seem to be age-related differences in the speed with which individuals can access the information in WM. Processing speed has been studied using the Sternberg task (Sternberg, 1966). In this task subjects are shown a set of items to be remembered, such as digits. On different trials the number of items in the set varies. On each trial, subjects are asked whether a target item is contained in the set. The time it takes for subjects to respond correctly (reaction time) as a function of the number of items in the set, is considered their WM processing speed. Older adults in the Sternberg task have significantly longer reaction times than young people (Anders, Fozard, & Lillyquist, 1972; Cerella, 1985; Eriksen, Hamlin, & Daye, 1973). WM processing speed for older adults is generally about one-half the speed of young adults (Kausler, 1991). The speed at which information is processed may play an important role in crash risk because information about potential traffic hazards must be thought about rapidly to avoid dangerous situations (West et al., 1992). Significant correlations have been found between "hesitancy" in decision making and crash rates among people over age 60 (West et al., 1992; French, West, Elander, & Wilding, 1993) and between lack of "thoroughness" in decision making and crash risk (Reason, Manstead, Stradling, Parker, & Baxter, 1991). Thus the age-related decline in processing speed may show up on the road as slow driving, hesitant driving, and unexpected maneuvers, all of which probably combine to increase the crash risk of older drivers.

Long-term memory (LTM) stores peoples' experiences and knowledge. All of the things that we know, and all that we are, are stored in LTM. It appears that the capacity of LTM is unlimited, or at least large (Tulving, 1974), and that it is possible for information to remain in LTM for a lifetime (Bahrick, Bahrick, & Wittlinger, 1975) regardless of age. There also does not seem to be much age difference in how fast information in LTM is searched (Thomas, Waugh, & Fozard, 1978). However, studies suggest that older adults have more difficulty than young adults transmitting information to LTM (Arenberg, 1976; Craik, 1968; Parkinson, Lindholm, & Inman, 1982; Salthouse, 1980). That is, older adults have more difficulty forming new LTMs than younger individuals. With changing road and vehicle characteristics, it is not surprising that drivers age 65 and older report having more difficulty driving now than in the past (Kostyniuk, Streff, & Eby, 1997). The ability to accurately retrieve information from LTM also seems to decline with age. Several studies have shown that when compared with younger individuals, older adults have greater difficulty with recalling information from LTM (Holland & Rabbitt, 1991; Kausler & Puckett, 1980, 1981; Kausler, 1990; Light, La Voie, Owens, & Mead, 1992). Thus even healthy older drivers may have greater difficulty than young drivers remembering what to do in certain driving situations and recalling driving laws, leading to increased crash risk.

PROBLEM SOLVING

The most complex cognitive activity that individuals engage in is problem solving, including decision making. Finding accurate and efficient solutions to problems is paramount to the task of driving. Older adults generally believe that their problem-solving ability improves with age (Williams, Denney, & Schadler, 1983). However, there is strong empirical evidence that problem-solving ability decreases with increasing age, with the quality of problem solutions decreasing after age 40–50 (Denney & Palmer, 1981; Denney & Pearce, 1989; Denney, Pearce, & Palmer, 1982). As reviewed in Kausler (1991), age-related declines in problem-solving ability have been found for numerous tasks, including number problems (Wright, 1981), the water jug problem (Heglin, 1956), the 20 questions paradigm (Denney & Denney, 1974), and reasoning (Cornelius & Caspi, 1987). Thus it appears that general problem-solving ability begins to decline after age 40–50, but the relationship between this decline and driving ability or traffic crash experience has yet to be determined.

SPATIAL COGNITION

Spatial cognition refers to the ability and knowledge to think about the arrangement of objects in space, including one's position relative to the environment (Matlin, 1989). Spatial cognition is used frequently in driving and may be related to the safe and efficient operation of a motor vehicle. Spatial abilities are used when drivers attempt to find their way across town and in solving other spatial problems. There is consistent evidence that spatial cognition ability declines with increasing age (Kausler,

1991; Ogden, 1990). Studies of mental rotation speed (i.e., the speed at which a person can rotate imagined images of objects in a mental rotation experiment; Shepard & Metzler, 1971) have shown that older adults are slower at rotating images than are young adults (Berg, Hertzog, & Hunt, 1982; Cerella, Poon, & Fozard, 1981; Gaylord & Marsh, 1975; Jacewicz & Hartley, 1979). In a study by Gaylord and Marsh (1975), older adults were approximately two times slower at mental rotation than young adults. In addition to slowed mental rotation times, some studies have shown that the accuracy of rotation declines with increasing age (Gaylord & Marsh, 1975; Herman & Bruce, 1983). Aubrey, Li, and Dobbs (1994) found that older adults have greater difficulty than young adults in understanding and using "you are here" maps like those found in shopping malls. Salthouse (1987) found that when compared with young adults, older adults had greater difficulty solving problems that required them to mentally integrate a series of lines to determine what they would look like if combined, called a *synthesis problem*. Collectively, these results suggest that older drivers take longer and have more difficulty performing tasks that require spatial thinking, which could lead to detrimental effects on traffic safety. However, the link between spatial cognition ability and driving performance has not been empirically determined.

Other abilities mediated by spatial cognition are cognitive mapping and way finding. Although the two abilities are highly related, *cognitive mapping* refers to the ability to accurately represent a spatial environment mentally (Matlin, 1989), and *way finding* refers to the ability to navigate efficiently in an environment. Cognitive mapping ability has been shown to decrease with increasing age. Older people are also less able than younger individuals to create detailed and organized mental maps of their neighborhoods, even when they have lived in their neighborhood an average of 18 years (Walsh, Krauss, & Regnier, 1981), and are likely to have more difficulty and be less accurate than young adults in remembering and recalling landmark locations in a simulated simple environment (Ohta, 1983). Navigation ability has also been shown to decline with increasing age, with older drivers reporting more difficulty navigating and finding locations than when they were younger (Kostyniuk et al., 1997a, 1997b) and exhibiting less navigational accuracy than young adults (Wochinger & Boehm-Davis, 1995).

CONCLUSIONS

The chapter has reviewed what is known about how psychomotor, visual, and cognitive abilities can decline with age. It is important to remember that the trends we have identified are for the population of older adults in general. An individual older adult may or may not be experiencing the declines we describe. In addition, some of these declines can be remediated through medical intervention and/or exercise. For example, a cloudy lens of the eye can be surgically replaced by a clear artificial lens greatly improving certain aspects of eyesight and a loss of flexibility can be improved through stretching and exercise. Finally, even though we discuss these declines separately, many can occur together, magnifying the potential effects of the declines on driving. The older driver who is concerned about his or her driving should see a medical or driving professional as described in Chapters 6 and 7 or seek more information (Chapter 9).

3 Critical Driving Skills

The one thing that unites all human beings, regardless of age, gender, religion, economic status or ethnic background, is that, deep down inside, we ALL believe that we are above average drivers.

Dave Barry

INTRODUCTION

Critical driving skills allow us to safely and efficiently operate a motor vehicle in traffic. These skills have to do with how we exert control over the vehicle; how we interact with other road users and perform various driving maneuvers such as yielding, turning, changing lanes, and passing; and how we make broader decisions about trip planning and finding our way along our route. As outlined in the previous chapter, age-related declines in psychomotor, visual, and cognitive abilities can adversely affect many critical driving skills. For example, declines in head/neck flexibility and/or peripheral vision can undermine critical driving skills related to merging into traffic and changing lanes (Suen & Mitchell, 1998). Thus, efforts to help older drivers maintain safe mobility need to be based on a thorough understanding of not only the abilities that can decline with age, but also the critical skills needed for driving that can be compromised. This chapter describes some of the approaches that have been taken to describe critical driving skills including driver task analysis (McKnight & Adams, 1970), the development of a hierarchical model for driving skills and control (Michon, 1985), and recent work to extend the hierarchical model to address the interplay between critical driving skills and motives (Keskinen, 2007). Also discussed in the chapter is work to identify a set of critical driving skills that are most salient to older drivers (Eby et al., in press). Finally, the implications of declines in these skills for efforts to maintain safe mobility are reviewed.

DRIVING TASK ANALYSIS

The delineation of critical driving skills has been an important part of broader efforts over the past several decades to understand driving behavior. Some of the early work in this area focused on identifying key driving tasks and the specific behaviors involved in the performance of each task. Using this task analysis approach, McKnight and Adams (1970) identified 45 major driving tasks with more than 1,700 elementary tasks. Collectively, these task descriptions constitute an exhaustive inventory of motor vehicle driving. An example of their results for Task 34, "lane changing," is provided in Figure 3.1 (McKnight & Adams, 1970).

```
34.1. Decides to change lanes
      34.1. Determines whether a lane is legally permissible
            34.1.1.1. Looks for regulatory signs prohibiting lane change well
            in advance of the maneuver
            34.1.1.2. Observes pavement markings
      34.1.2. Looks to the rear for approaching traffic in new lane
            34.1.2.1. Checks rear view mirror
                  34.1.2.1.1. Vehicles passing in new lane
                  34.1.2.1.2. Following vehicles closing fast from rear in
                  new lane
                  34.1.2.1.3. Following vehicles about to enter new lane
            34.1.2.2. Looks out window to check blind spot, moving head
            enough to see around blind spot
                  34.1.2.2.1. Varies speed of car very slightly to help bring
                  into view any vehicle traveling in the blind spot at exactly
                  the same speed as the car
            34.1.2.3. On multi-lane roads, look for vehicles about to enter new
            lane from the adjacent far lane
34.2. Prepares to change lanes
      34.2.1. Signals intention to change lanes by activating directional signal
      and/or employing appropriate hand signal
      34.2.2. Adjusts vehicle speed
            34.2.2.1. Accelerates if possible or maintains speed
34.3. Changes lanes
      34.3.1. If possible, waits a few seconds after signaling before beginning
      lane change
      34.3.2. Turns wheel and enters new lane
34.4. Completes lane change
      34.4.1. Positions vehicle in the center of new lane
      34.4.2. Cancels directional signal
      34.4.3. Adjusts vehicle to traffic flow in new lane
```

FIGURE 3.1 An example of the task analysis approach to inventory driver behaviors involved in lane changing (adapted from McKnight & Adams, 1970).

A HIERARCHICAL MODEL OF DRIVER SKILLS AND CONTROL

The driving task analysis by McKnight and Adams was an important starting point for understanding driver behavior. At the same time, its practical applications were somewhat limited, given its sheer size and its focus on the lower levels of driving skills. Expanding on this approach, Michon (1985) argued that a cognitive approach was needed in driver research, given the central role of cognition in human behavior in the transportation and mobility environment. Based on his view that "the most characteristic human component in this system is its behavior as an intelligent if not quite infallible problem solver," he divided the problem solving task of drivers into three levels of skills and control—strategic (planning), tactical (maneuvering), and operational (control)—as part of a hierarchical model of driver skills and control (Michon, 1979, pp. 488–489).

STRATEGIC LEVEL

The strategic level encompasses the general planning stage of a motor vehicle trip, with most decisions taking place before the trip even begins (Smiley, 2004). Strategic behavior includes high-level decisions about trip goals, mode of transit, driving route, circumstances under which to drive (e.g., time of day, weather conditions), and evaluation of the costs and risks involved, as well as broader issues such as whether to continue to drive and where to live in relation to destinations of choice (Michon, 1985; Smiley, 2004). High-level strategic decisions have implications for driving risk; for example, avoidance of adverse driving conditions by older drivers can be considered a risk compensatory behavior (Summala, 1996).

TACTICAL LEVEL

The tactical level has to do with the actual maneuvers drivers make in traffic in response to conditions in the driving environment at any given time. These include maneuvers such as obstacle avoidance, gap and headway acceptance, turning, and overtaking. Drivers must adapt their individual behavior to other road user's behavior and to the specific traffic situations that arise; thus, knowledge of traffic rules and behaving in accordance with these rules are important components of the skill set at this level (Berg, 2006). However, because driving is a self-paced task, there can be large amounts of variance in how these tasks are carried out (Summala, 1996).

OPERATIONAL LEVEL

Operational or vehicle control behavior has to do with the details of driving and includes such things as the method used to scan the roadway, the amplitude and frequency of steering movements, and the variation in speed (Smiley, 2004). Skills at this level need to be automated for the most part; otherwise, drivers would have considerable difficulty managing the tremendous flow of information they need to process and decide upon while driving (Berg, 2006).

Michon's model is a hierarchical model in that successes and failures at the higher levels affect the demands on skills at the lower levels (Hatakka, Keskinen, Gregersen, Glad, & Hernetkoski, 2002). At the same time, control in the hierarchy is not simply a top-down process; changes in lower levels also have effects on the whole system. Summala (1996) provided examples of top-down and bottom-up effects. In the case of the former, drivers may shift automatic control to conscious monitoring at the operational level if they are confronted with sudden changes in the environment or if they exceed safety margin thresholds. In the case of the latter, drivers may adjust their decision making at the strategic level in response to problems encountered at the tactical level—for example, if the configuration of a particular route makes maneuvering difficult, over time drivers may choose a different route.

Another characteristic of the hierarchical model is that different levels of decision making require different types of information (Ranney, 1994). For example, while strategic decision making can be largely memory driven, requiring little if any new information, tactical and operational decisions are based on the immediate driving environment and are therefore primarily data driven (Norman & Bobrow, 1975).

Michon's model was intended to improve upon earlier driving behavior models such as mechanistic models (Greenburg, 1959; Edie & Foot, 1960), adaptive control models (McRuer & Weir, 1969; Weir & McRuer, 1968), information flow models (Kidd & Laughery, 1964; Laughery, Andersen, & Kidd, 1967), and motivational models (Klebelsberg, 1977; Fuller, 1984; Taylor, 1964; Wilde, 1982; Wilde & Murdoch, 1982) that lacked or had only a limited cognitive framework. His model has been compared to Rasmussen's (1980, 1987) model of knowledge-based, rule-based, and skill-based behavior. As reviewed by Smiley (2004), Michon's strategic behaviors are comparable to knowledge-based behaviors, which involve conscious problem solving and are generally initiated in novel situations for which existing rules do not apply. Such decisions are made over time and not on a second-to-second basis. Tactical behaviors occur at the rule-based level at which choices are determined by the rules of the road and experiences in traffic. The rules allow some flexibility in driving behavior but choices are limited. Operational behaviors occur at the skill-based level, at which they are highly automated after much practice. Drivers are unconscious of the details of this behavior.

CRITICAL DRIVING SKILLS MOST SALIENT FOR OLDER DRIVERS

Taken together, all of the critical driving skills identified by Michon provide a useful context for understanding the driving behavior of individuals. In the case of older drivers, some of these critical driving skills are more salient than others in that they pose particular challenges for aging individuals. The driving skills needed to negotiate intersections can be especially problematic for older drivers as a result of declines in functional abilities (Eby, Molnar, & Pellerito, 2006; Oxley, Fildes, Corben, & Langford, 2006; Staplin, Gish, Decina, Lococo, & McKnight, 1998). Oxley et al. (2006) examined factors associated with intersection crashes among older drivers in Australia and found that the driving skill most compromised in crash-involved older drivers was selecting safe gaps in conflicting traffic when turning across or crossing traffic at intersections. Their study results indicated that improper gap selection was a factor in over 76 percent of the crashes and that gap selection was exacerbated by additional factors such as high vehicle speeds and traffic volumes.

Gap acceptance was also identified as a risk factor for older drivers in intersection crashes by Staplin, Lococo, McKnight, McKnight, and Odenheimer (1998). Based on results of an observational field study to document driving problems and errors of older drivers at intersections, they found that problems with this skill resulted in the highest error rate for drivers, as captured by camera. Other problematic driving skills identified in the study had to do with failure to stop completely at a stop sign, stopping over a stopping bar, improper turning path, and stopping for no reason. Two recent studies similarly found that older drivers involved in intersection crashes were more likely to fail to yield the right-of-way, disregard the traffic signal, be responsible for the crash, be at stop-controlled or uncontrolled intersections, and be turning left (Braitman, Kirley, Ferguson, & Chaudhary, 2007; Mayhew, Simpson, & Ferguson, 2006), indicating compromised driving skills related to intersection negotiation.

Not all driving skills that are problematic for older drivers relate to intersection negotiation. McKnight (1988) found three key areas of compromised driving skills including speed (misjudging speed, driving too slowly, excessive braking), search (inattention, inadequate scanning, failure to observe to the rear, pulling out without looking), and vehicle control (less precise visual control, including deficiencies in maintaining path, maintaining speed, changing lanes, coming to a smooth stop, backing, left turns, and right turns). His review also indicated that the most frequent citation for older drivers involved failure to heed stop signs, traffic lights, no-left-turn signals, and other signs and signals, with citations being attributed not to willful disregard but rather to failure to take notice. Older drivers were also found to be over-involved in certain types of crashes including improper yielding of right-of-way, left turns, backing and parking, and sudden slowing or stopping in the traffic stream.

Building on findings from this literature, as well as the opinion of experts involved in a recent project on older drivers, a consensus-based set of strategic, tactical, and operational skills considered to be most salient for older drivers was recently developed (Eby et al., in press). This combined set of critical driving skills is summarized in Table 3.1.

ADDING DRIVER MOTIVES INTO MICHON'S MODEL

Michon's hierarchical model of driver skills and control has recently been expanded to take into account the experiences and motivations of drivers, which interact with drivers' skills and therefore affect their driving performance (Keskinen, 2007). This work, represented by Hatakka et al. (2002), Keskinen (1996), Keskinen, Hatakka, Laapotti, Katila, and Peraaho (2004), and Laapotti and Keskinen (2004), among others, combine the performance aspects of driving behavior (highlighted by Michon) with the motivational and attitudinal aspects of driving behavior.

The model identifies four interacting levels of driving behavior. The first three levels, vehicle maneuvering, mastering traffic situations, and goals and contexts of driving, correspond directly to Michon's operational, tactical, and strategic levels, respectively. The fourth level, goals for life and skills for living, is new and has to do with drivers' general motives and attitudes in life and how they affect driving; this level is connected not only to the motives and personal development of drivers but also to the cultural norms of society (Laapotti & Keskinen, 2004). The premise underlying the fourth level is that factors related to what individuals are like and how they live their day-to-day lives also affect approaches to driving and specific driving behaviors (Berg, 2006). Among these factors are personality traits such as self-control, as well as lifestyle, social background, gender, age, and group affiliation (Gregersen & Berg, 1994; Hatakka, 1998; Jessor, 1987; Schulze, 1990).

The overall behavior of drivers is seen as the end result of continuous interaction between the various levels, although the higher levels override and direct the lower levels. The specific areas of knowledge and skills at each level are summarized in Table 3.2 (Hatakka et al., 2002). We consider the fourth level a "life goals" level for our purposes of understanding and addressing the safe mobility of older adults within Michon's framework.

TABLE 3.1
Critical Driving Skills Most Salient to Older Drivers

Critical Driving Skill	Definition
Pretrip planning	Pretrip planning involves all of those things that you can do to make driving easier and safer before you even start your vehicle, such as making sure you are well rested, ensuring that you are not impaired from the side-effects of medications, planning your trip route in advance, and wearing your seat belt.
Way finding	Way finding is your skill in being able to drive to places where you want to go without getting lost. A driver who is lost can be dangerous because he or she often makes mistakes while driving that can lead to traffic crashes.
Yielding	Yielding involves knowing which vehicles have the right-of-way and waiting until those vehicles have passed before pulling into the road. Proper yielding is important for safe driving because drivers of other vehicles will be expecting you to know who has the right-of-way.
Turning	Skill in turning the vehicle is a basic part of safe driving. Turning, however, involves more than just using the steering wheel; it also involves approaching and exiting turns at the correct speed.
Responding to traffic signals/signs	Traffic signs, signals, and pavement markers (such as stop signs) are designed to help traffic move efficiently and safely. They are effective only if people notice them, know what they mean, and respond to them appropriately.
Changing lanes	Changing lanes involves checking to make sure the traffic lane is free of other vehicles, signaling your intent to change lanes, and then steering the vehicle into the next lane. The most dangerous part of changing lanes is that a vehicle can be in your blind spot.
Passing	Passing another vehicle involves knowing where passing is legal, being able to judge the speed of oncoming vehicles, and being able to accelerate safely and drive your car around a slower moving vehicle that is in front of you.
Observing	An important skill for safe driving is to maintain awareness of what is happening around you. Observing involves both paying attention to what is happening in front of you and also using the vehicle's mirrors to keep track of what is happening behind you. A person with good observing skills is aware of how traffic is changing and can make appropriate driving adjustments to avoid potentially dangerous situations.
Gap acceptance	An important driving skill is to be able to pull into or across traffic only when there is a large enough "gap" in traffic so that you can safely complete your maneuver. The gap is the length of time in which there is no traffic crossing your intended path and is determined both by the distance between vehicles and by the speed they are traveling. Research shows that many crashes are caused by drivers inappropriately judging a gap length.

TABLE 3.1 (continued)
Critical Driving Skills Most Salient to Older Drivers

Critical Driving Skill	Definition
Speed	The appropriate driving speed is based on the posted speed limit and the conditions of the roadway. Speeds should be reduced, for example, if the road is slippery. Note that traffic crashes can be caused by drivers traveling too fast and by drivers traveling too slow.
Backing up	Backing up is an important driving skill because it is difficult to see what is behind your car and it can be hard to steer while traveling in reverse. This skill involves being able to use mirrors and being able to turn your neck and body to see what is behind the vehicle.
Maintaining proper lane position	The lines on the road are designed to show drivers where cars should be driven. Exceeding these lines can lead to traffic crashes. Proper lane position is important not only while traveling in a traffic lane but also while negotiating intersections.
Following	Maintaining a proper following distance behind the vehicle in front of you is important for the prevention of rear-end crashes. The proper following distance is determined by the distance between you and the next vehicle, as well as the speed at which you are traveling. With greater travel speeds, a larger distance between vehicles is needed for safe following.
Signaling	Signaling your intent to turn or change lanes is important for safe driving because it lets other drivers and pedestrians know what you are about to do. It is equally important to remember to turn off the signal after you turn or change lanes so that others know you are done.
Use of headlights	Headlights are important for being able to see while driving at night and during bad weather. They are also important for letting other drivers and pedestrians know where you are when seeing is difficult. Remembering to turn off the bright high beam for approaching vehicles is also important for safe driving because the high-beam lights can make seeing difficult for the other driver.

From Molnar & Eby (2008b).

Although the motivation for adding the fourth level of skills and control was originally to address the elevated crash risk of young drivers (Gregersen & Berg, 1994), the expanded model has important applications for older drivers, with regard to maintaining safe mobility. Older drivers experiencing declines in functional abilities may be able to extend safe driving by making adaptations to their driving behavior. However, such adaptations are generally either not possible or not effective at the lower levels of critical driving skills and control. For example, older drivers are unlikely to make operational adaptations, such as changing the frequency of their steering movement, given the automated nature of driving at this level. Furthermore, although tactical adaptations such as reducing driving speed or increasing headway may decrease the stress of driving for older drivers, these adaptations can actually

TABLE 3.2

Knowledge and Skills by Level of Driver Behavior

Level	Knowledge and Skills
Goals for life and skills for living (life goals)	Lifestyle/life situation
	Group norms
	Motives
	Self control and other characteristics
	Personal values
Goals for context of driving (strategic)	Effects of trip goals on driving
	Planning and choosing routes
	Evaluation of requested driving time
	Effects of social pressure in car
	Evaluation of necessity of trip
Mastery of traffic situations (tactical)	Traffic rules
	Observation/selection of signals
	Anticipation of course of situations
	Speed adjustment
	Communication
	Driving path
	Driving order
	Distance to others/safety margins
Vehicle maneuvering (operational)	Tire grip and friction
	Vehicle properties
	Physical phenomena

Adapted from Hatakka et al., 2002.

increase the risk of a crash in some circumstances as older drivers come into contact with younger drivers who tend to drive at faster speeds and tolerate shorter headways (Smiley, 2004).

The greatest opportunity for driving adaptations is at the higher levels of driving skills and control. For example, many older drivers make changes at the strategic level in terms of how much they drive and under what circumstances (e.g., time of day, weather conditions, type of road). These changes may be in response to the recognition that functional abilities have declined. They may also come about as a result of changes in social needs, often due to changing social roles brought about by major life changes such as retirement or the death of a spouse (especially a husband who has done the driving; Smiley, 2004). For these latter changes, the fourth level of skills (life goals) is particularly salient. For example, it is at this highest level that lifestyle decisions are made, such as what kind of motor vehicle to drive. Safety-conscious consumers tend to buy larger cars, which has important implications for crash risk (Summala, 1996).

CONCLUSIONS

Driver skills and control occur at various levels. Michon's model of strategic, tactical, and operational levels not only serves as a useful foundation for understanding the critical driving skills of older adults, but also provides a framework for thinking about the decisions that older adults make that affect driving and safe mobility. These decisions can be thought of as a form of behavioral adaptation or self-regulation. Adding the fourth level of "life goals" to Michon's model provides valuable additional insights into older adult behavior because it is often these larger motives, tendencies, and social relationships in the broader sense that affect individuals' goals and the context of driving (Berg, 2006). Thus, efforts to enhance the safe mobility of older drivers need to address these larger life goals behind driving and the context in which driving occurs.

4 Medical Conditions and Driving

You know you're getting old when all the names in your black book have M.D. after them.

Harrison Ford

INTRODUCTION

An important theme of this book is that aging per se does not lead to declining abilities that are needed for safe mobility. Rather, there are a number of medical conditions and other factors that are more likely to occur in the older adult population that increase risk and are, therefore, associated with aging.[*] In fact, it is not the condition itself that raises the risk of a crash, but rather how the condition influences functional abilities—those abilities that underlie critical driving skills. A fully managed medical condition, such as high blood pressure, may not affect driving at all. This chapter reviews what is known about how specific medical conditions affect crash risk or, in the absence of risk data, functional abilities. We focus here on chronic (enduring, predictable, and stable) rather than acute (short-term, sporadic, and unpredictable) effects (Dobbs, 2005). During acute events (e.g., cardiac infarction or seizure) individuals are not capable of driving safely and should not do so.

Crash risk is influenced by many factors. As discussed by Dobbs (2005) and Charlton et al. (2004), not all medical conditions influence crash risk to the same degree and individuals with the same condition differ greatly in how they are affected. Individuals also vary in their abilities and resources to compensate for declining functional abilities. In addition, some chronic conditions can involve acute episodes, drastically increasing crash risk and making judgments of fitness to drive difficult given a specific condition. Many chronic medical conditions are progressive, such as Parkinson's disease. A medical condition may not affect driving safety in its early stages, but may compromise driving safety in later stages. Finally, people often have comorbid medical conditions, that is, more than one medical condition. The combined effects of the conditions make it difficult to determine the effects of a specific medical condition on driving performance and crash risk.

There are a variety of conditions that can cause declines in critical driving skills. Here we classify them into several general categories: visual perception, auditory

---

[*] The medications used to treat these conditions can also compromise safe mobility and are discussed in Chapter 5.

perception, cardiovascular, cerebrovascular, neurological, metabolic, respiratory, musculoskeletal, and psychiatric. Within each category we review those conditions that are either age-related or common in the older adult population.

VISUAL PERCEPTION

Seeing well is of obvious importance for safe driving. Indeed, the information needed for driving is predominately visual (Sivak, 1996) and all U.S. jurisdictions require vision testing for licensure. There is clear evidence that many changes in visual function are related to aging. Deficits in visual function can impact a wide range of critical driving skills. Here we review several age-related visual problems and how they influence driving performance and crash risk.

CATARACTS

A cataract is the clouding of the normally clear lens of the eye (National Eye Institute, NEI, 2008a). Cataracts are quite common in the older U.S. population, affecting about one-half of people age 80 and older (NEI, 2008a). Cataracts affect vision by blocking some of the light entering the eye and can lead to reduced acuity, sensitivity, night vision, color vision, and size of the visual field. Cataracts can be present in one or both eyes.

Drivers with cataracts report difficulties with driving more often than those without cataracts (Owsley et al., 1999) and they self-regulate their driving in many ways including avoiding nighttime driving, inclement weather, expressway travel, high-traffic volume areas and times, and left turns, reducing total driving exposure, and quitting driving altogether (Ball, Owsley, Stalvey et al., 1998; Marottoli et al., 1993; Owsley et al., 1999). This self-regulation of driving suggests that drivers are aware of their visual declines and adjust driving accordingly, but also that many people are in fact driving with cataracts. This is surprising, since cataract treatment is highly effective and safe. According to the National Eye Institute (2008a), cataract removal with intraocular lens implantation surgery improves vision in 90 percent of cases.

Untreated cataracts can increase crash risk even with driver self-regulation. Work by Owsley and colleagues found that drivers with cataracts were 2.5 to 6 times more likely to be in an at-fault crash than drivers with no eye disease (Owsley et al., 1999; Owsley & McGwin, 1999; Owsley, Stalvey, Wells, Sloane, & McGwin, 2001). These relative risks were adjusted for driving exposure, age, comorbid conditions, and several other variables that could affect crash risk, independent of cataracts. Fortunately, cataract surgery can significantly reduce crash risk. Owsley et al. (2002) found that the crash rate for drivers in the five years following cataract surgery was one-half that of a matched group of drivers with cataracts. Postsurgery drivers also reported less difficulty driving after the surgery.

COLOR VISION DEFICITS

Color vision/discrimination is the ability to perceptually distinguish among different wavelengths of light. Color deficiency is present in about 8 percent of males and

1 percent of females in the United States, with the prevalence slightly higher in the older adult population (Prevent Blindness America, 2005). As discussed by Owsley and McGwin (1999), several states test for color discrimination abilities during license renewal, presumably to assess whether or not a person can distinguish the colors of a traffic control signal. Even for a person who is color deficient, there is other information that can be used for perceiving the state of traffic signals, including relative location and luminance. Studies have shown that color deficiency does not result in compromised driving ability or increased risk of a crash (Ball, Owsley, Sloane, Roeneker, & Bruni, 1993; Owsley, Ball, Sloane, Roenker, & Bruni, 1991; Owsley & McGwin, 1999; Vingrys & Cole, 1988).

DIABETIC RETINOPATHY

According to the National Eye Institute (2008b), diabetic retinopathy is the leading cause of blindness in Americans. This diabetic eye disease causes changes in the blood vessels of the retina, the light-sensitive layer of cells in the back of the eye needed for vision. People with diabetic retinopathy may have blood vessels that bleed or leak, or have abnormal blood vessels that easily rupture obscuring vision (NEI, 2008b). Treatment is more effective in the early stages of the disease. Because there are no symptoms in the early stages, a comprehensive eye exam is the only way to discover blood vessel problems. In the late stage of the disease, called *proliferative retinopathy*, vision becomes blurry and blind spots in the central visual field can develop (NEI, 2008b). Treatment in the early stages of the disease can be effective in preventing blindness in 90 percent of cases (NEI, 2008b).

Despite the clear effects on visual functioning, few studies have addressed the impact of diabetic retinopathy on driving and crashes. In one population-based case-control study in Washington State, the crash histories of drivers with several visual conditions, including diabetic retinopathy, were compared to matched control drivers (McCloskey, Koepsell, Wolf, & Buchner, 1994). The study found that drivers with retinopathy had a slightly *lower* risk of a crash than control drivers, although this difference was not statistically significant. There were no data reported on the adaptive behaviors of drivers with retinopathy, so these results could not be corrected for possible differences in exposure to a crash. Clearly more research is needed in this area.

GLAUCOMA

Glaucoma is the set of eye diseases characterized by increased pressure inside the eye (NEI, 2008c). Over time, the rise in pressure can damage the optic nerve resulting in vision loss and blindness. People with glaucoma gradually lose vision in the periphery (NEI, 2008c). Glaucoma is the leading cause of blindness for African Americans and the second leading cause of blindness for all Americans, affecting about 3 million Americans over age 40 (National Study to Prevent Blindness, 1980; Patlak, 1990; Wilson, 1989). It is estimated that about one-half of people with glaucoma are unaware that they have the disease (NEI, 2008c). One small study in Australia found that a sample of drivers with glaucoma rated their vision as "good" even though 85 percent of the group had peripheral vision loss (Carberry, Wood, & Watson, 2004).

People with visual field loss resulting from glaucoma report that driving is more difficult (Gutierrez, Wilson, & Johnson, 1997; Parrish et al., 1997) and self-regulate driving by limiting night, freeway, and unfamiliar area driving (Adler, Bauer, Rottunda, & Kuskowski, 2005). Evidence on the relationship between glaucoma and crash risk is mixed. A study in Alabama found that when compared to a control group with no vision problems, a group of glaucoma patients was 3.6 times more likely to have been in a crash in the previous 5 years. A similar study in Canada with a smaller sample of people found that drivers with glaucoma were 6 times more likely to have been in a crash in the previous 5 years and were 12 times more likely to have been at fault (Haymes, LeBlanc, Nicolela, Chiasson, & Chauhan, 2007). Other studies have found glaucoma to be a risk factor in self-reported crashes (Foley, Wallace, & Eberhard, 1995; Szlyk, Mahler, Seiple, Edward, & Wilensky, 2005) and simulator crashes (Szlyk et al., 2005). At the same time, other work has found no significant relationship between glaucoma and increased crash risk (McCloskey et al., 1994; McGwin, 2004; Owsley, McGwin, Mays et al., 2004; Stewart, Moore, Marks, May, & Hale, 1993; Szlyk, Taglia, Paliga, Edward, & Wilensky, 2002) and one study found a decreased crash risk for drivers with glaucoma (Owsley, McGwin, Mays et al., 2004).

These mixed results may be due in part to the progressive nature of glaucoma. Although the disease has no cure, if detected early, loss of further visual function can be slowed or prevented. Those whose disease is detected early will generally have a less significant loss of the visual field. When the amount of visual field loss due to glaucoma is related to crash risk, moderate or severe field loss is associated with an increased risk for involvement in a motor vehicle crash, while low visual field loss is not (McGwin et al., 2005).

Monocular Vision

Monocular vision refers to vision in only one eye, regardless of what causes the loss of vision. The two main visual deficits caused by monocular vision are a reduction in the visual field and a loss of depth cues that require vision from both eyes (binocular). Early work on monocular vision and driving found that monocular drivers had as great as two times the crash rate as the general population, engaged in more hazardous driving, and had more reckless driving violations (Freytag & Sachs, 1969; Keeney, 1968; Keeney, Garvey, & Brunker, 1981). This work, however, was limited by small sample sizes, potentially unrepresentative samples, and a failure to take into account the driving exposure and other factors that could be unique to monocular drivers.

Recent work has found that monocular drivers perform as well as binocular drivers during on-road assessment (McKnight Shinar, & Hilburn, 1991; Racette & Casson, 2005) and in a driving simulator (Wood & Troutbeck, 1994). One author has suggested that monocular race car drivers could safely compete, pointing out, "...we have proof that it is possible for the one eyed individual to participate as a racing driver without major incident for at least one season." (Westlake, 2001,

p. 622). Indeed, when researchers take into account the known changes in driving behavior for drivers with visual field loss (e.g., Coeckelbergh, Brouwer, Cornelissen, Woffelaar, & Kooijman, 2002; Freeman, Muñoz, Turano, & West, 2006), the crash and violation rates for monocular and binocular drivers seem to be the same (Johnson & Keltner, 1983; Rubin et al., 2007).

Age-Related Macular Degeneration (ARMD)

ARMD is a progressive disease that destroys the light-sensitive cells in the central part of the retina, known as the *macula* (NEI, 2008d). The macula is the part of the retina that is used for sharp, central vision—the type of vision needed for reading, many driving tasks, and other activities of daily living (Scilley et al., 2002). People with ARMD have difficulty resolving sharp detail in their central vision. The largest risk factor for ARMD is age, followed by being female, Caucasian, and a smoker, and family history (Dobbs, 2005; Charlton et al., 2004). There are no consistently effective treatments for ARMD and the prevalence is estimated at about 9 percent of the population age 65 and older (Charlton et al., 2004), with great regional differences around the world (Klein, Klein, & Cruickshanks, 1999).

A number of studies have found that drivers with ARMD engage in many driving behavioral adaptations, such as driving more slowly, avoiding nighttime driving, and avoiding challenging driving situations (Ball, Owsley, Stalvey et al., 1998; DeCarlo, Scilley, Wells, & Owsley, 2003; Moore & Miller, 2005; Jackson, Owsley, & Cideciyan, 1997; Mangione, Ajani, & Padan, 1994; Szlyk et al., 1995). Despite the prevalence and the visual disability caused by ARMD, there have been few studies looking at the relationship between this condition and crash risk. Two studies have examined the effects of ARMD on driving performance and self-reported crash risk (Szlyk, Fishman, Severing, Alexander, & Viana, 1993; Szlyk et al., 1995). Both studies had small samples and found no differences between the ARMD drivers and a control group, except for nighttime crashes. Although many of the ARMD drivers reported no nighttime driving, those that did were more likely to be in nighttime crashes. These studies did not correct for driving exposure. With the known behavioral adaptations for ARMD drivers, these crash rates are difficult to understand.

A more recent study with a large sample found that drivers with ARMD had an injury-crash risk three times that of control drivers (Owsley, McGwin et al., 1998). Interestingly, no significant difference was found for noninjury crashes. Further, the authors found that injury-crash involvement for ARMD drivers was partially related to other exposure and disease variables, weakening the findings.

Collectively, the results of these studies suggest four conclusions: (1) people with ARMD drive; (2) those who do drive greatly limit their driving, particularly at night; (3) these driving adaptations are generally protective—they seem to prevent crashes; and (4) those ARMD drivers who choose to drive at night probably should not.

AUDITORY PERCEPTION

It seems that the ability to sense and process auditory information is important for driving, yet there is little research to support this notion. There are numerous tests for hearing sensitivity, both objective and subjective. Hearing sensitivity has been measured in many studies addressing the relationship between cognitive and perceptual measures and measures of at-risk driving (Kantor, Mauger, Richardson, & Unroe, 2004; Marottoli, Cooney, Wagner, Doucette, & Tinetti, 1994; McCloskey et al., 1994; Sims, McGwin, Allman, Ball, & Owsley, 2000; Sims, Owsley, Allman, Ball, & Smoot, 1998; Szlyk, Myers, Zhang, Wetzel, & Shapiro, 2002). Most of these studies do not report the relationship between hearing ability and unsafe driving, presumably because there was no significant relationship found. In those studies that have reported hearing outcomes, no significant relationship was found between hearing loss and police-reported injury crashes or between hearing loss and simulated driving performance. One study, however, indicated that drivers using hearing aids had an increased injurious crash risk, possibly due to the distraction from the hearing aides' feedback (McCloskey, Koepsell et al., 1994).

CARDIOVASCULAR

The cardiovascular or circulatory system consists of the heart and blood vessels that carry oxygen and nutrients to the cells of the body and remove carbon dioxide and other waste products. There are a number of age-related conditions that affect the functioning of the cardiovascular system that can adversely impact safe driving.

CORONARY HEART DISEASE

Coronary heart disease (CHD) is caused by a restriction or disruption of the arteries that supply the heart with blood, so that the heart does not have the oxygen and nutrients needed to properly pump blood. The risk factors for developing CHD include smoking, high blood pressure, high cholesterol, diabetes, obesity, physical inactivity, and stress (Mayo Clinic, 2006). CHD is relatively uncommon for young people but is the leading cause of death among U.S. individuals age 65 and older (Centers for Disease Control and Prevention, CDC, 2004). Although CHD is a chronic and progressive condition, slightly more than 40 percent of deaths from CHD occur during a sudden acute event (Kannel, Gagnon, & Cupples, 1990).

The incidence of death from this condition while driving, however, is thought to be quite low, because drivers experiencing an acute event (e.g., heart attack) are generally able to stop their vehicles before crashing (Epstein et al., 1996; Janke, 2001; Schmidt, Haarhoff, & Bonte, 1990). Studies relating CHD and crashes have generally found that CHD did not increase the risk of a crash and may have even reduced the risk (Charlton et al., 2004; Dobbs, 2005; Janke, 1994). These findings, however, do not indicate that those with CHD are safer drivers. Most likely, the reduced risk of a crash is related to changes in driving behavior such as reduced driving or driving during safer conditions. Thus, there does not seem to be a serious problem with heart disease and driving safety among older drivers, because those with heart disease seem to appropriately self-regulate their driving activities.

ARRHYTHMIA

Arrhythmia is an irregular rhythm of the heart. The condition itself does not seem to negatively affect driving ability (for reviews, see Dobbs, 2005; Charlton et al., 2004), but the treatment can have serious consequences. A common treatment for the condition is a pacemaker or implantable cardioverter-defibrillator (ICD). ICDs are used to manage arrhythmia by delivering a high-energy electric shock to the heart to restore proper rhythm. This shock can sometimes result in loss of consciousness (syncope) or temporary impairment of movement (Kou et al., 1991; Epstein et al., 1996). The incidence of loss of consciousness after shock delivery can be as high as 15 to 21 percent in those with ICDs (Axtell & Akhtar, 1990; Bansch et al., 1998; Fogoros, Elson, & Bonnet, 1989; Grimm, Flores, & Marchlinski, 1993; Kou et al., 1991; Maloney et al., 1991; Ruppel et al., 1998). It is difficult to predict when and if a shock will lead to loss of consciousness or paralysis. There are no clinical predictors for loss of consciousness or paralysis related to arrhythmia, and even a patient's history of these side effects or their absence does not predict future occurrences (Dobbs, 2005; Kou et al., 1991). Thus, it is difficult to determine whether patients with ICD-treated arrhythmia should drive. Recent research, however, suggests that well-treated arrhythmia should not preclude a person from driving (Bleakley & Akiyama, 2003).

CONGESTIVE HEART FAILURE

Congestive heart failure (CHF) occurs when the heart does not provide enough blood flow to suit the body's needs. The causes of CHF include disease of the coronary artery, heart attack, damaged heart valves and/or muscle, arrhythmia, and congenital birth defects (Mayo Clinic, 2008). The condition causes blood to congest in various organs of the body, leading to organ dysfunction or failure. People with CHF are often fatigued; have shortness of breath and swelling of arms, legs, and/or body; lack appetite; and have difficulty remaining alert (Mayo Clinic, 2008). CHF is more common in older individuals. According to the National Institutes of Health (1996), about 1.7 percent of the U.S. population (4.8 million people) have CHF, while the incidence for those age 70 and older is 10 percent. Unfortunately, there has been no work relating CHF to driving performance or crash risk.

ABNORMAL BLOOD PRESSURE

Abnormal blood pressure conditions include hypertension (high blood pressure) and hypotension (low blood pressure). Hypertension is quite common, affecting about 30 percent of Americans (Hajjar, Kotchen, & Kotchen, 2006). An estimated 19 percent of people age 60 and older with hypertension are unaware of the problem or are untreated (Ong et al., 2007). There has been little research on the effect of hypertension on crash risk. It is known, however, that hypertension can lead to other chronic conditions such as a stroke, coronary heart disease, a heart attack, or dementia (Dobbs, 2005). These conditions can have a serious effect on traffic safety as described elsewhere in this chapter.

Hypotension is less common than hypertension, affecting an estimated 10–20 percent of older adults (WebMD, 2008). Chronic hypotension is generally not considered a problem, except when blood pressure abruptly drops causing lightheadedness or syncope (WebMD, 2008). Again, this chronic condition has not been studied in relation to traffic safety. Prudence suggests, however, that individuals who experience frequent syncopal episodes (fainting) should avoid driving until a medical professional indicates that driving can be safely undertaken.

CEREBROVASCULAR

Cerebrovascular conditions are those that cause a lack of blood flow to a part of the brain, such as a clot in a brain blood vessel, a rupture of a vessel, or a change in blood viscosity (National Institute of Neurological Disorders and Stroke, 1990). The loss of blood flow to a certain part of the brain causes cells to die, disrupting the functioning of that part of the brain. The specific effects of the condition are related to where in the brain the damage occurs and how much of the brain is involved. Common effects are paralysis, difficulty speaking, vision problems, weakness, and balance problems (Lings & Jensen, 1991). There are two main age-related cerebrovascular diseases that could affect driving: transient ischemic attack and stroke. Each type of disease starts with an acute event. The lasting effects of the event are chronic.

TRANSIENT ISCHEMIC ATTACK

A transient ischemic attack (TIA) is an acute loss of certain brain functions with symptoms lasting anywhere from several minutes to a day (Warlow & Morris, 1982). As with all kinds of cerebrovascular events, the symptoms vary greatly and depend on the location of the event in the brain. Because permanent cerebral damage does not generally occur with a TIA, driving should not be affected once the symptoms have dissipated. It is important, however, to seek medical attention after a TIA as it is a warning sign for a much more serious stroke. According to some research, of those who have had a first TIA, 30 percent will experience a stroke in the following 3 years (Mohr & Pessin, 1986).

STROKE

Stroke, or cerebrovascular accident (CVA), becomes more likely as a person ages. Unlike TIAs, the symptoms of CVA are more severe and long lasting. The prevalence rates for the 65 and older age group are as much as 10 times higher than for the overall population (Kurtzke, 1985). The physical effects of a stroke depend largely on the location in the brain where the stroke occurred, but these effects can include partial or incomplete paralysis, impaired visuospatial abilities, agnosias, aphasia, attention deficits, impaired recognition ability, reduced numerical ability, and emotional disruptions (Lings & Jensen, 1991). Research shows that 30–42 percent of stroke survivors return to driving, many without receiving advice about their fitness to drive (Fisk, Owsley, & Pulley, 1997; Legh-Smith, Wade, & Hower, 1986). Although there is a growing body of literature on the assessment of stroke patients for driving fitness (e.g., Klavora, Heslegrave, & Young, 2000; Lundberg, Caneman, Samuelsson, Hakamies-Blomqvist, & Almkvist, 2003), there is little research on the link between stroke and traffic crashes. Of the work that has been done, methodological issues make interpretation of the results in terms of stroke-related crash risk difficult (Diller et al., 1998; Haselkorn, Mueller, & Rivara, 1998). There are no universally accepted criteria for assessing fitness to drive after a stroke. However, because the effects of strokes are individual, driving rehabilitation specialists, occupational therapy generalists, and other health-care professionals should help stroke patients in making decisions about their driving capability, including when it is prudent to reduce driving or cease driving altogether.

NEUROLOGICAL

Neurological disorders are diseases that affect the functioning of the central, autonomic, and peripheral nervous systems. There are hundreds of neurological conditions. Here, we review what is known about the more common age-related disorders, as well as some common disorders that affect the aging population but are not more likely with increasing age.

SYNCOPE

Syncope can result from a variety of causes including a sudden drop in blood pressure, a neurological pathology, and an increase in blood sugar (Rehm & Ross, 1995). It occurs most frequently in the older population, with a prevalence of about 3 percent among those age 65 and older (Bonema & Maddens, 1992; Kapoor, 1994; Savage et al., 1985). In about 40 percent of cases of syncope, no cause can be found (Kapoor et al., 1983; Kapoor, Hammill, & Gersh, 1989; Spudis, Penry, & Gibson, 1986). The chance that a person who has had a previous episode of syncope will have another acute episode while driving is low. Research has shown that of syncope patients studied, 9 to 12 percent of them had one or more syncopal events while driving (Li et al., 2000; MacMahon, O'Neill, & Kenny, 1996). A syncopal episode while driving, however, is serious (Li et al., 2000; Sheldon & Koshman, 1995). Li et al. (2000), for example, found that in syncope-related crashes about 37 percent resulted in injury and about 5 percent resulted in a fatality.

There is little agreement about whether syncope patients should drive, with recommended guidelines ranging from complete driving cessation of 1 to 3 months after a single episode of syncope, to driving cessation for 1 year, to complete cessation after multiple episodes of syncope (Canadian Cardiovascular Society, 1996; Decter, Goldner, & Cohen, 1994; Sheldon & Koshman, 1995).

EPILEPSY/SEIZURES

Epilepsy is a chronic neurological condition that occasionally causes abnormal electrical activity in the brain, resulting in seizures (Epilepsy Foundation, 2008). Seizures can range from dramatic grand mal (also known as *tonic-clonic*) seizures to subtle seizures, all of which can lead to adverse changes in cognition and/or consciousness. While the cause of epilepsy is unknown in about three-quarters of cases, risk factors include vascular disease, head trauma, congenital factors, central nervous system infections, and neoplasms (Hauser & Kurland, 1975). Epilepsy is relatively common in the United States, with prevalence estimated at 5–7 people per 1000 in the general population, and slightly lower in the older adult population (CDC, 1994). Typical treatment involves drug therapy, vagus nerve stimulation, changes in diet, and, in severe cases, surgery (Mayo Clinic, 2007).

The main risk of epilepsy for driving is that the occurrence of a seizure can cause loss of consciousness and motor control. Because seizures often occur without warning, drivers may not have enough time to safely stop their vehicle. This potential for crashes is supported by studies that have found an increased risk of crashes and injury among drivers with epilepsy (Diller et al., 1998; Hansotia & Broste, 1991; Popkin & Waller, 1989; Waller, 1965). Those with epilepsy should drive only after seeking the advice of a driving professional, such as a certified driving rehabilitation specialist.

Sleep Apnea

Sleep apnea is characterized by snoring, breath cessations, sleep disturbances, and daytime drowsiness (Haraldsson, Carenfelt, & Tingvall, 1992). The condition has been shown to affect various abilities related to safe driving such as forced choice and delayed reaction times, decreased vigilance and attentive abilities, impaired cognitive functioning, and psychomotor difficulties (Bédard et al., 1991; Findley et al., 1986; Greenberg, Watson, & Depula, 1987; Kales et al., 1985). One obvious concern of apnea is drowsiness while driving. Studies have found that up to 54 percent of people with untreated apnea report falling asleep while driving and having drowsiness-related "near crashes" or actual crashes, compared to only 7 percent of matched nonapnea control drivers (Engleman et al., 1996; Guilleminault, Van den Hoed, & Mitler, 1978; Gonzalez-Rothi, Foresman, & Block, 1988). In addition, drivers with untreated apnea perform significantly worse than matched controls on simulated driving tests (Findley et al., 1989; Haraldsson et al., 1992; George, Boudreau, & Smiley, 1996). When compared to controls, apnea drivers have an increased risk of crash of 2 to 15 percent (Barbé et al., 1998; Findley et al., 1988; George et al., 1987; George & Smiley, 2001; Horstman et al., 2000; Terán-Santos, Jiménez-Gómez, & Cordero-Guevara, 1999; Young et al., 1997). Apnea drivers are also overrepresented in single-vehicle but not multiple-vehicle crashes (Haraldsson, Carenfelt, & Tingvall, 1992) and have significantly more at-fault crashes and traffic citations than drivers without apnea (Findley, Unverzagt, & Suratt, 1988). Thus, it is clear that apnea patients should be evaluated for driving fitness.

Multiple Sclerosis

Multiple sclerosis (MS) is an autoimmune disease of the central nervous system (brain, spinal cord, and optic nerves) characterized by inflammation and destruction of the myelin sheath surrounding nerve cells (National Multiple Sclerosis Society,

2006). The myelin is necessary for normal nerve function. The symptoms of the disease are intermittent and variable affecting different people in different ways and even the same person in different ways, depending on which nerve cells are being damaged. According to the National MS Society (2005), about 400,000 Americans have MS, and women are two to three times more likely to have MS than men.

MS has the potential to affect nearly all functioning of the body, depending on where the inflammation is in the central nervous system. Symptoms can include numbness, tingling, weakness, paralysis, spasms, blindness, blurred or double vision, loss of balance, fatigue, depression, short-term memory deficits, decreased information processing speed, and many other types of cognitive dysfunction (Jones, 2004). Many of these symptoms affect safe driving, in particular the cognitive symptoms (Lincoln & Radford, 2008). Indeed, a simulator study found that MS drivers performed more poorly than controls in simulated driving designed to measure lane keeping under high cognitive load and car following (Marcotte et al., 2005). In this study, MS drivers drove at higher speeds, had difficulty matching the speed of a lead vehicle, and had difficulty staying in their lane. Other work has also established that cognitive impairment resulting from MS degrades driving performance (Schultheis, Garay, & DeLuca, 2001).

In addition to problems with driving performance, drivers with MS also seem to be at an increased risk of a crash. Studies in Europe have found that drivers with MS have higher numbers of traffic citations and crashes than do matched healthy drivers (Knecht, 1977; Schanke, Grismo, & Sundet, 1995; both cited in Schultheis, Garay, & DeLuca, 2001). Recent work in the United States has also found an increased crash risk for drivers with MS, but only for those MS drivers with cognitive impairment (Schultheis et al., 2002; Shawaryn et al., 2002). These studies found no increase in citations among MS drivers.

Determining whether an MS patient should drive can be difficult. While impairments of psychomotor ability are of obvious importance (Goodwill, 1984), the emotional problems and cognitive deficits are, perhaps, more important factors to consider (Schanke, Grismo, & Sundet, 1995; Schultheis et al., 2001). A driver with MS should be evaluated by a driving professional to determine his or her fitness to drive.

PARKINSON'S DISEASE

Parkinson's disease (PD) is a brain disorder that affects nerve cells that produce the vital neurotransmitter known as *dopamine* (National Parkinson's Foundation, 2008). Dopamine is necessary for the brain to smoothly control movement. When about 80 percent of the dopamine-producing cells have died, the symptoms of PD appear. These symptoms include tremors, slowed movement, stiffness, and poor balance (National Parkinson's Foundation, 2008). PD also causes cognitive impairment including memory deficits, slowed information processing, decreased sustained and divided attention abilities, and decreased visuospatial awareness (Radford, Lincoln, & Lennox, 2004). The National Institute of Neurological Disorders and Stroke (2004) estimates that there are about 500,000 cases of PD in the United States, with 50,000 new cases reported each year. The average age of onset is 60 and both the prevalence and incidence of PD increase with age. At present there is no cure, although symptoms can be lessened by drug treatment.

PD is a progressive disorder, with symptoms gradually worsening over the course of many years. Because of the progressive nature of PD, drivers with this disorder will need to give up driving at some point, but may have difficulty knowing when that time has arrived (Campbell, Bush, & Hale, 1993). Research in the United Kingdom comparing PD drivers' cognitive and physical deficits with their on-road driving performance found that cognitive abilities were not associated with fitness to drive, but that physical declines were (Radford et al., 2004). Other work, however, in which PD drivers were given secondary or distracting tasks, found that when the cognitive load is high, PD drivers made more errors than matched controls during simulated (Stolwyk, Triggs et al., 2006; Stolwyk, Charlton et al., 2006) and on-road driving (Uc et al., 2006a,b). Thus, it appears that both the psychomotor and cognitive symptoms of PD can affect driving safety.

Indeed, a number of studies have shown that the driving abilities of people with PD are compromised (Devos et al., 2007; Heikkilä, Turkka, Kallanranta, & Summala, 1998; Singh, Pentland, Hunter, & Provan, 2006; Wood, Worringham, Mallon, & Silburn, 2004; Zesiewicz et al., 2002). Despite the demonstrated effect of PD on driving ability, there is a paucity of studies that have examined crash risk. Of the three studies that have investigated this issue, drivers with PD were found to be slightly more at risk than healthy controls and those drivers with more severe symptoms had higher crash rates (Dubinski et al., 1991; Meindorfer et al., 2005; Ritter & Steinberg, 1979, cited in Meindorfer et al., 2005).

Frucht et al. (1999) reported a new side effect of PD drug treatment—sudden onset of sleep (SOS). The authors presented eight case studies in which PD patients fell asleep while driving and crashed. Five of these drivers claimed that there were no warning signs (e.g., drowsiness) prior to falling sleep. While sleepiness is a known side effect of PD drug treatment, there is some controversy about whether SOS is a real phenomenon. As described by Meindorfer et al. (2005), it is possible that SOS episodes are preceded by drowsiness and misjudged or forgotten by the person. Meindorfer et al., and others (Hobson et al., 2002) concluded that if real, SOS episodes are rare.

DEMENTIA

Dementia/Alzheimer's (DA) is a fatal disease characterized by intellectual deterioration, particularly memory loss, in an adult that is severe enough to interfere with occupational or social performance (McKhann et al., 1984). This condition occurs almost exclusively in the older adult population. DA can be caused by a variety of medical conditions including stroke, hypothyroidism, acquired brain injuries, brain tumors, carbon monoxide poisoning, and alcoholism (Haase, 1977; Katzman, 1987). Because of variation in how DA is diagnosed, prevalence estimates range from 4 to 16 percent of the older adult population (Adler, Rottunda, & Dusken, 1996; Cushman, 1992; Evans et al., 1989; Terry & Katzman, 1983). Three severity stages of DA have been indexed by the Clinical Dementia Rating Scale: early, middle, and late (Hughes et al., 1982). Progression usually spans an average of 8 years from the time symptoms first appear, although DA has been known to last as long as 25 years. People with early-stage dementia do drive and studies show that up to 45 percent of all DA patients still drive (Carr, Jackson, & Alguire, 1990; Logsdon, Teri, & Larson, 1992; Lucas-Blaustein et al., 1988), with the vast majority of these people driving alone (Lucas-Blaustein et al., 1988). Perhaps because of impaired insight, research shows that people with DA do not change their behaviors after a crash (Lucas-Blaustein et al., 1988).

Research has plainly shown that individuals with dementia drive more poorly than drivers without dementia. Studies have identified several driving problems associated with DA, including getting lost while driving, even in familiar areas (Adler et al., 1996; Lucas-Blaustein et al., 1988; Silverstein, Flaherty, & Tobin, 2002; Underwood, 1992); vehicle speed control difficulties (Odenheimer et al., 1994), particularly driving consistently below posted speed limits (Lucas-Blaustein et al., 1988); failure to signal lane changes (Hunt et al., 1993; Odenheimer et al., 1994); failure to check blind spots before lane changes (Hunt, et al., 1993); failure to maintain lateral lane position (Odenheimer et al., 1994); running stop signs (Cushman, 1992); and failure to recognize and obey traffic signs (Adler, Rottunda, & Dusken, 1996; Carr et al., 1998; Cushman, 1992; Hunt et al., 1993; Mitchell, Castleden, & Fanthome, 1995; Uc et al., 2005). As DA progresses, these errors appear to become more frequent (Fox et al., 1997). Indeed, in a comprehensive review of the literature, Man-Son-Hing, Marshall, Molnar, and Wilson (2007) reviewed 17 case-control studies that examined the driving abilities in DA drivers using on-road assessment, driving simulators, and reports from caregivers. The review found that drivers with DA performed worse than control drivers in all 17 studies.

Such findings suggest that drivers with dementia would have a higher crash rate than would cognitively intact drivers. Research evidence, however, is inconclusive. Early research reported elevated crash risk for drivers with DA (e.g., Drachman & Swearer, 1993; Dubinski, Williamson, Gray, & Glatt, 1992). A more recent analysis by Man-Son-Hing and colleagues (2007) review six case-control studies that investigated DA and crash risk using caregiver-reported and state-recorded crashes. All three studies that based their conclusions on caregiver-reported crash information found that DA patients had more crashes than nondemented drivers. As we have described elsewhere in this book, self-reported crash data can be limited and biased. Of the three studies utilizing objective state-recorded crash data, only one found that DA patients had more crashes. The other two studies found no differences. More research is needed to understand how the poor driving performance in DA patients might or might not translate into an elevated crash risk.

METABOLIC

Metabolic disorders cause a disruption in the body's ability to produce energy within cells. Of the many metabolic disorders, here we focus on two of the most common: diabetes and thyroid disease.

Diabetes Mellitus

According to the American Diabetes Association (2008), a person is diabetic when his or her body does not produce or properly use insulin, a hormone that helps to convert food into energy for cells. Diabetes is thought to be genetic, with obesity and an inactive lifestyle contributing to its likelihood. It is estimated that about 7 percent of the U.S. population has diabetes and it is more common in the older population (20.9 percent for those age 60 and older; American Diabetes Association, 2008).

Diabetes is classified into two types: Type 1 (insulin dependent) and Type 2 (non-insulin dependent). Five to 10 percent of all people diagnosed with diabetes have Type 1 diabetes and the remaining have Type 2 diabetes (CDC, 1997). Diabetes causes a variety of vascular problems that can lead to various health conditions including heart attacks, visual deficits (see section on diabetic retinopathy), and loss of feeling in the extremities, which can in turn directly affect the ability to drive safely. Insulin and other medications used to control diabetes can also adversely affect driving abilities. Other effects of diabetes occur when blood glucose (energy) levels are incorrect. If the levels are too high (hyperglycemia), a person can have visual and mental problems and life-threatening acute events (Charlton et al., 2004). If glucose levels are too low (hypoglycemia), a person's body will first respond with an autonomic response (tremor, sweating, hunger, anxiety, rapid heart rate). If the person does not take action to restore the glucose level, the person will begin to experience cognitive, physical, and visual declines (Dobbs, 2005). All of these symptoms can affect the ability to drive safely.

A simulator study where glucose levels of Type 1 diabetes patients were experimentally manipulated found significant declines in driving abilities for moderate hypoglycemia (Cox et al., 2001). These declines included more swerving, slow driving, and lane keeping difficulty. These results were consistent with previous work by the authors (Cox, Gonder-Frederick, & Clark, 1993).

Studies relating crash risk and diabetes have yielded inconsistent results (Charlton et al., 2004; Dobbs, 2005; Janke, 1994). Using a variety of subject groups and methods for collecting crash-involvement information, some studies reported that diabetic drivers were at increased risk of crash (Hansotia & Broste, 1991; Laberge-Nadeau et al., 2000; Songer et al., 1988), other studies found no difference between diabetics and other groups (Kennedy et al., 2002; McGwin, Sims, Pulley, & Roseman, 1999; Stevens et al., 1989), and one study found a lower crash risk for diabetic drivers (Eadington & Frier, 1989). Differences in the populations, reporting methods, measures of driving exposure, and the symptomology of the disease may account for these discrepancies. Clearly, more research is needed in this area.

Of recent interest is the potential for a hypoglycemic reaction while driving. This acute event could cause serious driving problems. Cox et al. (2001) measured glucose levels during a simulated drive and found that the intrinsic metabolic demands of driving can lower glucose to hypoglycemic levels. Other work has shown that some drivers are not able to accurately perceive when glucose levels drop (Clark, Cox, Gonder-Frederick, & Kovatchev, 1999). Several studies found that self-reported hypoglycemic reactions were involved in 15 to 60 percent of crashes with diabetic drivers (Chantelau, 1991; Eadington & Frier, 1989; Frier, Matthews, Steel, & Duncan,

1980; Steel, Frier, Young, & Duncan, 1981; Stevens et al., 1989). Thus, current advice is to educate diabetic drivers about the warning signs of hypoglycemic reactions and to have them monitor glucose levels prior to driving (Clarke et al., 1999; Graveling, Warren, & Frier, 2004; Harsch et al., 2002).

Thyroid (Hypothyroidism)

Hypothyroidism is a condition in which the thyroid gland does not produce enough thyroid hormone (EndocrineWeb, 2005). The condition is more prevalent in women (about 9 percent) than men (about 1 percent; Vanderpump et al., 1995) and is more likely to occur as one ages (Laurberg et al., 1999). Among other symptoms, the condition can cause fatigue, sleepiness, and cognitive impairment. If untreated, the disease can cause sleep apnea, dementia, heart failure, and other serious conditions that can compromise driving safety. Fortunately, with thyroid replacement therapy, symptoms generally disappear in a few weeks, although there is some evidence suggesting that cognitive symptoms may be permanent. Properly treated hypothyroidism should not affect driving safety.

RESPIRATORY

The respiratory system is composed of the organs that allow one to breathe so that oxygen can be delivered to the body and carbon dioxide removed. Diseases of this system cause the body to be deprived of an adequate level of oxygen and cause a buildup of excess carbon dioxide. Here we focus on two of the common respiratory disorders: asthma and chronic obstructive pulmonary disorder.

Asthma

Asthma occurs when the air passages for breathing become inflamed causing them to temporarily constrict or become clogged. Asthma symptoms can be triggered by a number of causes including allergens (e.g., dust mites), irritants (e.g., cigarette smoke), and medicines. The symptoms of asthma include difficulty breathing, coughing, chest tightness, and wheezing. The symptoms can be so severe that death can result from a lack of oxygen (NIH, 2008). Asthma is most common in people under age 18 (75/1,000), but it still affects about 40 in every 1,000 people age 65 and older in the United States (NIH, 1999). Asthma is more deadly in the elderly population—60 percent of all asthma deaths in the United States are people age 65 and older (Asthma and Allergy Foundation of America, 2008). There are no published studies that have addressed the effects of asthma on driving performance or crashes.

Chronic Obstructive Pulmonary Disease

Chronic obstructive pulmonary disease is a class of conditions that disrupt the functioning of the respiratory system. The two most common COPDs are emphysema and chronic bronchitis (COPD International, 2004). There are an estimated 16 million cases of COPD in the United States and the prevalence increases with age. COPD is the 4th leading cause of death for people age 65–84 in the United States (COPD International, 2004). The symptoms of COPD are similar to asthma, but much more severe and long lasting. The symptom of greatest concern is chronic hypoxemia, or lack of oxygen in the blood. As reviewed by Dobbs (2005), chronic hypoxemia can lead to cognitive decline, especially when performing demanding and complex tasks.

As with asthma, there are few studies examining COPD and driving performance or crash risk. One retrospective case control study examined crash rates and "pulmonary conditions," including COPD and other conditions (Vernon et al., 2001). The study found a small but significant increase in crash risk for drivers with pulmonary conditions. This study, however, did not control for driving exposure, so the crash rates may not be meaningful. At the same time, given that driving is a complex and demanding activity, it seems prudent that COPD patients be regularly screened for cognitive decline and counseled about fitness to drive.

MUSCULOSKELETAL

The musculoskeletal system enables the body to move. The system is composed of the muscles, bones, joints, tendons, and ligaments. There are multiple conditions that can affect the functioning of this system, all of which make it harder or more painful to move some or all of the parts of the body. Here we focus only on arthritis.

ARTHRITIS

Arthritis is a type of musculoskeletal disease characterized by inflammation, pain, and stiffness in or around the joints of the body (CDC, 2006). There are numerous conditions that cause arthritis and the location and severity vary greatly (CDC, 2006). More than 46 million people in the United States have doctor-diagnosed arthritis and this number is expected to grow to 67 million by 2030 as the baby boomers age. Arthritis is more common in the older age groups with 50 percent of people age 65 and older having some form of doctor-diagnosed arthritis (CDC, 2006).

Given the physical nature of the driving task (e.g., steering, braking, accelerating), it is not surprising that arthritis can adversely affect driving performance. Research has found that people with rheumatoid arthritis (an autoimmune disorder that attacks the joints) had difficulty performing many driving tasks, including steering, cornering, using a parking brake, and reversing (Jones, McCann, & Lassere, 1991). A study in Connecticut found that among community-living individuals age 72 and older, those with foot impairments (including arthritis) were more likely to have self-reported crashes, citations, and police contacts than those without these impairments (Marottoli et al., 1994). Other work examined the relationship between arthritis and motor vehicle crashes by considering crash records, finding that those with a diagnosis of arthritis were about three times more likely to be in a crash (Tuokko, Beattie, Tallman, & Cooper, 1995). This work also found that arthritic patients taking non-steroidal anti-inflammatory drugs (NSAIDS) had an even higher risk of crash, suggesting that the treatment may also impair driving abilities. Drivers with arthritis should consult a driving professional to determine fitness to drive.

PSYCHIATRIC

Psychiatric disorders are thoughts or behavioral patterns that are abnormal for an individual's development or culture. There are several hundred psychiatric disorders, the vast majority of which are not related to aging but can influence driving (see Charlton et al., 2004, or Dobbs, 2005, for excellent reviews). Here we review one disorder, depression, which is prevalent in the older adult population.

DEPRESSION

Many people feel depressed from time to time. Such depression is usually short lived and is often an appropriate response to an adverse life condition. However, clinical depression is long term and is characterized by chronic feelings of worthlessness and sadness, loss of interest in things that used to be pleasurable, loss of energy, disturbed sleep, loss of appetite, thoughts of suicide, and difficulty concentrating (WebMD, 2006). As discussed in a report by the U.S. Department of Health and Human Services (USDHHS, 1999), the prevalence of depression in the older adult population is difficult to estimate. This report cites estimates ranging from 5 to 35 percent for adults age 65 and older. There are several reasons why the prevalence is difficult to estimate. First, the definition and procedures for diagnosis vary widely (USDHHS, 1999). Second, older patients are reluctant to report depression symptoms to their physician (USDHHS, 1999). Third, depression is frequently coupled with other medical problems such as diabetes complications (DeGroot et al., 2001) and macular degeneration (Brody et al., 2000). Thus, it is difficult to disentangle the effects of a comorbid condition from depression.

Silverstone (1998) reports that clinically depressed people are at high risk of a crash because of slowed reaction times, lack of concentration leading to distracted driving, and suicidal or self-mutilation thoughts, which may lead to risky driving or even a deliberate crash (see also, Rubinsztein & Lawton, 1995). Indeed, although the research is sparse, some studies show that depressed people are at higher risk of crash. In a 5-year prospective study, Sims et al. (2000) found that clinically depressed older adults (age 55 and older) were 2.5 times more likely to be in a crash than non-depressed adults. Foley, Wallace, and Eberhard (1995) found a similar result among older drivers in a rural community. On the other hand, a study of crash risk factors among older women did not find an association between depression and crashes (Margolis et al., 2002). There is a need for further research in this area.

CONCLUSIONS

We have reviewed the significant literature on what is known about common age-related medical conditions and driving. We have also presented some of the difficulties with relating medical conditions to declines in driving performance and increased crash risk. As you may have concluded from this chapter, there are few clear-cut cases in which older adults with medical conditions can be ruled either safe or unsafe to drive simply on the basis of the medical condition. For older adults with a medical condition, the decision to drive in most cases should be made based on the advice of a physician and, if necessary, a driving professional and a family member. As we emphasize in several places in this book, each individual is unique, and these personal variables should be taken into account when analyzing a person's transportation safety and mobility.

In this chapter we have treated each condition separately, although many people have comorbid conditions. Very little research has considered the combined effects of multiple medical conditions or how various treatment options affect traffic safety and mobility. These issues are fertile areas for future research.

5 Medications and Driving

You can't help getting older, but you don't have to get old.

George Burns

INTRODUCTION

Medicine has been a part of civilization since ancient times. 3,500-year-old texts from Ancient Egypt describe remedies to cure a variety of ailments (Dollinger, 2007). For example, the ancient Egyptian Ebers Papyrus describes giving an infusion of willow bark as a pain reliever. In the mid-1800s, chemists isolated an ingredient in willow bark and called it *salicylic acid*, which the Germans later refined into a pill they named Aspirin (Bellis, 2008a). The German company, Bayer, patented Aspirin in 1900—the same medicine that can still be purchased worldwide today. After the development of vaccines for childhood diseases in the 1950s and 1960s, the development of new drugs accelerated exponentially, resulting in a modern and vast pharmacopoeia today.

The relationship among aging, medication use, and motor vehicle crashes is extremely complex. While medications can impair driving for any age group, older drivers in particular may be overly affected by medication use. As pointed out by Wilkinson and Moskowitz (2001), older people are more likely to use multiple medications (polypharmacy), have adverse changes in pharmacokinetics (how the body metabolizes a medicinal compound), and experience adverse changes in pharmacodynamics (how the body reacts to a medicinal compound). Among older adults, there may also be inappropriate use of medications, underutilization of medications, and overuse of medications (Wilkinson & Moskowitz, 2001). Finally, in addition to prescription medication, older adults may consume over-the-counter medications, herbal supplements, and alcohol. Polypharmacy, pharmacokinetics, pharmacodynamics, the medical conditions for which the medications are being used, and the timing and strength of a dose all complicate the task of relating traffic crashes to medication use. In addition, as we have discussed, older drivers are known to regulate their own driving, reducing crash risk by simply reducing their exposure to crashes.

Our formulation of this issue is shown in Figure 5.1. As shown, driver behavior is influenced by individual characteristics, health conditions, and the medications taken to treat those conditions. Individual characteristics can be numerous and include a driver's tolerance for risk, behavioral adaptations, preferences, and driving history. These individual characteristics have a great influence over driving skills and make the study of medication use and driving complicated because it

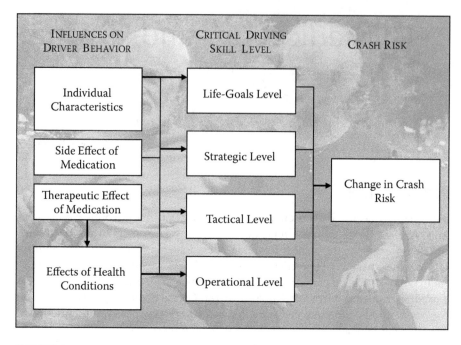

FIGURE 5.1   Illustration of how the therapeutic and side effects of medications, as well as other influences on driver behavior, affect critical driving skills and crash risk.

is difficult to separate the effects of a drug from the effects of individual characteristics on driving skills. Also as shown in this figure, medications have both a primary (therapeutic) effect on a health condition and side effects that can directly influence critical driving skills. A driver's health conditions, whether treated or untreated, also influence critical driving skills directly. Properly treated conditions can improve driving skills, while untreated or mistreated conditions can have deleterious effects on driving skills.

Recognition that medications might affect driving comes fairly recently. The majority of automobile drivers in all developed Western nations occasionally or continually use medications that have the potential to impair the ability to operate an automobile. Recognition of this threat to drivers was not widespread before the late 1960s. Attention was called to this issue after the introduction of benzodiazepine tranquilizers and hypnotic drugs. For the first time, drugs designed to alter mood and behavior were used by tens of millions of fully ambulatory patients (O'Hanlon & De Gier, 1986). Many of these mood-altering drugs are called *psychotropic drugs*, because they affect the central nervous system (CNS). Drugs that affect the CNS also affect driving because the CNS mediates all sensation, most movement, and all thinking (Aparasu & Mort, 2004), all of which are required for safe driving.

Understanding the use of psychotropic drugs and their effects on driving is particularly important in the older driver population because psychotropic medications are most often used by persons age 65 and older (Leveille et al., 1994); psychotropic agents are inappropriately prescribed in up to one-third of older adult patients (Aparasu & Mort, 2004); and age-related declines in psychomotor function, vision, attention, information processing, and motor coordination may increase vulnerability to CNS effects of drugs (Ray, Thapa, & Shorr, 1993). Furthermore, Ray, Fought et al., (1992) and Ray, Gurwitz et al., (1992) found that drug use does not differ between groups of people over age 65 with and without driver's licenses, suggesting that medications with the potential to impair driving ability are not avoided by people who can legally drive.

This chapter will first review the prevalence of medication use in the older adult population, including polypharmacy, inappropriate prescription use, and nonprescription medicine use. We will then review what is known about the effects of medications and driving. Because the number of drugs is vast and constantly changing, here we review five of the most common drug classes.

PREVALENCE OF DRUG USE AMONG OLDER ADULTS

PRESCRIPTION MEDICATIONS

The use of medications increases with age. A U.S. study of noninstitutionalized older adults (Gurwitz, 2004) found that more than 90 percent take at least one medication per week. Gurwitz (2004) also found that polypharmacy was common with more than 40 percent of older adults using five or more prescription drugs per week and 12 percent using 10 or more drugs per week.

Prevalence of specific prescription drug use has been investigated in several studies (Cadigan, Magaziner, & Fedder, 1989; Chrischilles et al., 1990; Gurwitz et al., 2003; Leveille et al., 1994; Ray et al., 1993). In the most comprehensive cohort study to date, Gurwitz et al. (2003) examined the medical records and prescription drug use of more than 30,000 Medicare+Choice Plan enrollees for a period of 1 year. All people in the study were age 65 and older.

Table 5.1 shows the results of the Gurwitz et al. (2003) study as well as the results of other prevalence studies. As shown in Table 5.1, the five most frequently prescribed medications for older adults are each prescribed to more than 20 percent of the population, with cardiovascular medicines being by far the most frequently used by older adults.

INAPPROPRIATE PRESCRIPTION MEDICATION USE

Not all prescription medications are appropriate for older persons. Beers (1997) developed a list of "potentially inappropriate" medications for older people based on the following two criteria: (1) the drugs or drug classes that are ineffective or pose an unnecessarily high risk, or the drugs for which there is a safer alternative for people age 65 and older; and (2) medications that should not be used by older adults who have certain medical conditions. This list was later updated by Zhan et al. (2001), based on the review of an expert panel, who identified three categories of potentially inappropriate drugs for older adults: (1) those that should always be avoided; (2)

TABLE 5.1
Prevalence of Prescription Medications in the Older Adult Population

Medication Category	Percent of Older Adult Population Taking the Medication
Cardiovascular	53.2–55.8[1–3]
Antibiotic/anti-infective	5.2–44.2[1,2]
Diuretic	27.2 –36.4[1,2,6]
Opioids	21.7[1]
Antihyperlipidemic	21.7[1]
Nonopioid analgesic	19.8[1]
Gastrointestinal track	8.7–19.0[1,2]
Respiratory track	5.8–15.6[1,2]
Dermatologic	14.8[1]
Antidepressant	3.8–13.2[1,2,4,7]
Sedative/hypnotic	3.8–13.3[1,2,5]
Hypoglycemic	5.7–11.5[1,2,8]
Steroids	4.7–9.7[1,2]
Ophthalmics	3.7–9.6[1,2]
Thyroid	4.6–12.1[1,2,6]
Antihistamines	2.7–9.2[1,2,7]
Hormones	9.1–14.6[1,2]
Anticoagulants	2.7–7.0[1,2]
Muscle relaxants	5.5[1]
Osteoporosis	5.3[1]
Antiseizure	0.9–3.4[1,2]
Antigout	1.3–3.2[1,2]
Antineoplastics	2.8[1]
Antiplatelets	1.3[1]
Antipsychotics	0.7–1.2[1,2]
Antiparkinsonians	0.9[1]
Alzheimer's disease	0.9[1]
Immunomodulators	0.04[1]

[1] Gurwitz et al. (2003)
[2] Cadigan, Magaziner, & Fedder (1989)
[3] Helling et al. (1987)
[4] Blazer, Hybels, Simonsick, & Hanlon (2000a)
[5] Blazer, Hybels, Simonsick, & Hanlon (2000b)
[6] Stuck et al. (1994)
[7] Leveille et al. (1994)
[8] Ray, Thapa, & Shorr (1993)

those that are rarely appropriate; and (3) those that have indications but are often misused. Zhan et al. (2001) examined responses from a nationally representative sample of older adults from the 1996 Medical Expenditure Panel Survey. The results showed that 2.6 percent were taking drugs from the "always avoid" list; 9.1 percent were prescribed drugs from the "rarely appropriate" list; and 13.3 percent reported prescriptions from the "often misused" list. In all, 21.3 percent were prescribed drugs from one of the three lists. Zhan et al. (2001) concluded that "inappropriate use of medications in elderly patients remains a significant problem in the United States" (p. 2827). A number of other studies have documented the problem of inappropriate prescription use among older adults. Lococo and Staplin (2006) provide an excellent summary of this work.

NONPRESCRIPTION MEDICATIONS

Not included in Table 5.1 are prevalence estimates of nonprescription drugs and dietary supplement use. During the last decade, alternative medicines and dietary supplements have become more popular (Eisenberg et al., 1993, 1998). These medications can be complex mixtures of active ingredients that in many cases have unknown biological effects (Nahin et al., 2006). Also, because they are not subjected to the same regulations as prescription medications, the levels and mixtures of ingredients can vary greatly. Given these uncertainties, it is surprising that a large proportion of the U.S. population views these products as safe (Blendon et al., 2001) and only about one-half of individuals discuss their use with a medical professional (Bruno & Ellis, 2005).

Because these substances can be purchased over the counter, it is difficult to obtain objective measures of use in the older adult population. To date, studies have relied on self-reported data through face-to-face or telephone interviews. Two recent studies investigated supplement use among older adults (Fennel, 2004; Nahin et al., 2006). Fennel (2004) examined data from the 2000 National Health Interview Survey, a face-to-face nationally representative sample of noninstitutionalized U.S. citizens, to determine supplement use for 3129 adults age 70 and older. Nahin et al. (2006) analyzed data from a cross-sectional study of 3,072 ambulatory adults age 75 and older from sites in three states. Table 5.2 shows the results of these studies. Vitamins are by far the most commonly used supplement, and herbal medications are used by a small but growing minority.

A study by Bruno and Ellis (2005) examined data from the 2002 National Health Interview Survey to determine use of herbal remedies in older adults. The study found that 12.9 percent had used an herbal remedy at least once in the past year and that use was most prevalent in the 65–69 age group, among females, and among those with higher income and educational levels. Of those who reported use of herbal remedies, the 12 most commonly reported were glucosamine (23.2 percent), Echinacea (22.7 percent), garlic (21.9 percent), ginkgo biloba (15.3 percent), fish oils (11.8 percent), ginseng (9.7 percent), ginger (9.5 percent), saw palmetto (9.3 percent), soy (7.1 percent), peppermint (6.9 percent), St. John's wort (6.6 percent), and chamomile/ragweed (5.9 percent).

TABLE 5.2

Prevalence of Dietary Supplements and Herbal Remedies in the Older Adult Population as Found in Two Studies

Supplement	Nahin et al. (2006)	Fennel (2004)
Multivitamin	59.4	47.3
Vitamin A/Beta carotene	8.3	5.6
Vitamin B	17.8	*
Vitamin C	31.6	23.0
Vitamin D	22.4	*
Vitamin E	43.8	27.4
Calcium	37.2	*
Magnesium	7.3	*
Zinc	7.5	*
Ginseng	1.6	*
Garlic	4.0	*
Ginkgo	3.8	*
St. John's wort	0.4	0.8
Echinacea	2.3	2.0
Coenzyme Q10	2.2	*
Glucosamine	17.9	*
Saw Palmetto	3.6	*
Astragalus	*	0.2
Dong quia	*	0.2
Cat's claw	*	0.2
Any vitamin/mineral	66.6	60.6

* Not reported.

MEDICATIONS AND DRIVING SAFETY

Here we review what is known about how the major drug classes influence driving performance and crash risk. Any medicinal compound that causes drowsiness, memory problems, dizziness, movement difficulties, or visual problems can impair a person's ability to drive safely. As depicted in Figure 5.1, medicines and their side effects can adversely affect critical driving skills, which in turn can influence the risk of a crash.

BENZODIAZEPINES

Benzodiazepines are the class of medications that are CNS depressants. These are used to treat anxiety, muscle spasms, insomnia, and seizures, and as a sedative (Jones, Shinar, & Walsh, 2003; Ray, Thapa, & Shorr, 1993). These drugs are more commonly known as *tranquilizers* and *hypnotics*. After absorption, benzodiazepines are absorbed into the blood and the highly vascular tissues such as the brain, heart, lung, and liver, and also the less vascular peripheral muscles. The concentration of the drug in the body is greatly affected by the dosage level (Nicholson, 1986).

Concentration is also affected by how long the body takes to eliminate the drug. There are two kinds of benzodiazepines: "long half-life" and "short half-life." An evening dose of a long half-life hypnotic can markedly impair psychomotor function the next day, but a similar dose of a short half-life drug results in less impairment (Ray et al., 1993). While long half-life drugs generally are eliminated from the body in 24 hours, for older adults this elimination may take more than 72 hours because of an age-related decrease in metabolic efficiency (Regestein, 1992; Salzman, 1992).

Commonly prescribed benzodiazepines with long half-lives include diazepam (better known as *Valium*), flurazepam, and lorazepam. All have consistently been found to impair driving performance, especially during the morning after ingestion and sometimes into the afternoon (O'Hanlon et al., 1982; Moskowitz & Smiley, 1982). O'Hanlon et al. (1982) noted, however, that it may be possible for patients to adapt to long-term treatment of benzodiazepine, which may reduce the drug's effect on driving performance.

The relationship between benzodiazepine use and impairment of driving performance has been thoroughly studied (see Berghaus & Grass, 1997; Ray et al., 1993; Wilkinson & Moskowitz, 2001). The more than 500 experimental studies on this topic show consistent results: Driving performance declines with increased levels of benzodiazepines. Although short half-life benzodiazepines are eliminated faster, they can still be dangerous. O'Hanlon (1992) cited several studies that have found that short half-life benzodiazepines impair driving performance. The benefit of short half-life benzodiazepines is that the blood level decreases faster than the long half-life variety.

Unlike the literature on driving performance, the literature investigating the associations between benzodiazepines and crash risk is more limited. Studies with young drivers have generally shown that benzodiazepines increase the risk of a crash (summarized in Ray et al., 1993). Studies with older drivers have found similar results (Hemmelgarn et al., 1997; McGwin et al., 2000; Neutel, 1995; Ray et al., 1992; Sims et al., 2000). A study in the United States, for example, found a 50 percent increase in crash risk among older people who used benzodiazepines (Ray, Fought et al., 1992). A Canadian study found that within the first week of using benzodiazapines, there was a 45 percent increased crash risk among older adults who took long half-life benzodiazepines (Hemmelgarn et al., 1997). Crash risk decreased after the first week of use, but still remained 26 percent higher than controls, even after a year of continuous use. Interestingly, for older adults who took short half-life benzodiazepines, no increased crash risk was found. A study of older adult at-fault crash risk in Alabama found that when compared to control drivers, those taking benzodiazepines were more than five times as likely to be in an at-fault crash (McGwin et al., 2000). Collectively, this body of literature suggests two conclusions: (1) Older adults taking benzodiazepines are at a higher risk of a crash and should avoid driving for several hours after dosage; and (2) for older adults who drive, only short half-life benzodiazepines should be prescribed, if possible.

Antihistamines

Antihistamines are drugs used to alleviate the symptoms of mild allergic reactions. This class of drugs is widely available over the counter in the form of tablets, ointments, drops, and sprays. Some antihistamines produce drowsiness, sedation, or dizziness and are sometimes used to treat insomnia and motion sickness (Sisk, 2006).

Research shows that the sleep-inducing and sedative properties of this class of drugs may impair driving ability (Verster & Volkerts, 2004). One of the most common antihistamines is Benedryl® (diphenhydramine). This medication is available in "regular" and in nondrowsy forms. Research on regular diphenhydramine has found that it impairs driving performance, although the impairment lasted for only a 2–3 hour period after taking the medication (Gengo, Gabos, & Miller, 1989; Nakra, Gfeller, & Hassan, 1992; Ray et al., 1993). Nondrowsy/sedative antihistamines do not impair cognitive or psychomotor performance (Nicholson, 1986) and do not appear to affect the ability to drive an automobile (Betts et al., 1984)

The relationship between drowsiness/sedation-producing antihistamines and crash risk has not been firmly established. Studies that have attempted to determine the effect of this type of antihistamine on crash risk among older drivers have found that past users showed a slightly increased risk for an injurious motor-vehicle crash, as compared with nonusers (Leveille et al., 1994). However, antihistamines were not found to increase crash risk among the older adults in another study (Ray, Fought et al., 1992). Collectively, it is safe to conclude that older adults who drive should (1) use only the nondrowsy formulations of antihistamines; and (2) avoid driving for several hours after taking a regular antihistamine if it is needed.

Antidepressants

Antidepressants are prescribed for severe and clinical depression. There are three main subclasses of antidepressants: serotonin-specific reuptake inhibitors (SSRIs; e.g., Prozac®), monoamine oxidase inhibitors (MAOs; e.g., Celexa®), and tricyclic antidepressants (e.g., Nardil®; Internet Drug News, Inc., 2007). There are also newer antidepressants that inhibit the uptake of both serotonin and norepinephrine (e.g., Effexor®). Antidepressants can also be classified by whether or not they produce sedation as a side effect. Susceptibility to sedation as a side effect may increase with age (Sanders, 1986).

Use of sedating antidepressants is consistently associated with deterioration in a wide variety of vehicle-handling skills (Clayton, Harvey, & Betts, 1977; Hindmarch, 1988; Ramaekers, 2003; Seppala et al., 1975; Smiley, 1987). For example, utilizing an in-traffic driving test, sedating antidepressants significantly impaired lateral position control and speed control compared to a placebo (Louwerens, Brookhuis, & O'Hanlon, 1984). Indeed, in this study, several antidepressant-dosed participants did not complete their drives due to safety concerns, whereas all of the placebo-dosed participants completed the study. Several authors have reviewed the literature and concluded that sedating antidepressants can impair driving performance, while the nonsedating type does not, except possibly at high doses (Lococo & Staplin, 2006; Ray et al., 1993; Walsh et al., 2004; Wilkinson & Moskowitz, 2001).

Studies have also found that sedating antidepressants increase crash risk among older adults (Leveille et al., 1994; Ray et al., 1992). In two studies, older adults with current prescriptions of sedating tricyclic antidepressants had more than twice the risk for injury crashes, compared to older adults without these prescriptions (Leveille et al., 1994; Ray, Fought et al., 1992). A strong dose-response effect has also been found, with crash risk increasing as dosage levels of the drug increased. Older drivers with high-dose prescriptions had nearly three times the risk for injury crashes, compared with nonusers (Leveille et al., 1994). Another study of older drivers found that the crash risk associated with taking 125 mg of amitriptyline (a popular sedating tricyclic antidepressant) daily resulted in five times the crash risk associated with a dosage of 25 mg of amitriptyline (Ray, Fought et al., 1992).

The conclusions for antidepressant use and driving are similar to those for antihistamines. Older adults who drive should (1) use nonsedating antidepressants if they are effective; (2) avoid driving if they are drowsy, even if they are taking nonsedating antidepressants; and (3) work with their health-care provider to determine their fitness to drive.

Opioid Analgesics

Opioids are the class of drugs made from the natural and synthetic derivatives of opium. Opioids are CNS depressants and are used primarily for the treatment of chronic pain and coughing (Ray et al., 1993). The familiar codeine cough syrup is included in this class. The main driving-related problem with opioids is that they can cause sedation and psychomotor declines even in the less potent forms, such as codeine (Ray et al., 1993).

As described by Lococo and Staplin (2006), it is generally accepted that any driver who is beginning treatment with an opioid should avoid driving, as there is clear evidence of driving performance declines. In an experiment with a driving simulator, young subjects ingesting a single dose of 50 mg codeine had an increased number of driving-off-the-road episodes and more collisions than control subjects (Linnoila & Hakkinen, 1973). Once a person has reached a dosage and is accustomed to the drug's effects, the effects on driving and safety diminish. In a review of 48 studies of opioid-dependent/tolerant patients and driving-related skills, Fishbain et al. (2003) concluded that (1) there was consistent evidence for no impairment in driving performance as measured by driving simulators, off-road driving, or on-road driving; and (2) there was strong and consistent evidence that opioid-tolerant patients were not at a higher risk of citation or traffic crashes when compared to control groups. The authors provide useful driving recommendations for patients on opioid treatment.

- After beginning opioid treatment, or after a dose increase, do not drive for 4–5 days
- Do not drive if you feel sedated
- Report sedation effects to your physician
- Do not combine alcohol or illicit drugs with the opioids
- Avoid taking antihistamines while being treated with opioids

- Do not make any changes to your medicine regime without consulting your physician

ANTIDIABETIC MEDICATIONS

Antidiabetic medications include insulin and oral hypoglycemics and are used in the treatment of diabetes mellitus. These drugs are used to maintain blood glucose at acceptable levels (American Diabetes Association, 2008). This class of drugs has no known effects on driving ability or increased crash risk (Ray et al., 1993). However, as we have described in Chapter 4, low levels of glucose (hypoglycemia) can produce significant declines in driving performance and may increase crash risk. Since antidiabetic medicines are designed to lower blood glucose levels, diabetics should not drive until they are familiar with the dosage effects of their medicines.

CONCLUSIONS

In this chapter, we have reviewed the current knowledge related to the prevalence of medication and supplement use and the effects of medications on driving. In doing so, we have emphasized the difficulty of establishing the relationship between medications and crash risk. One of the most challenging issues in this field is polypharmacy. As noted by Carr (2004), "there are a myriad of sedating medications in this era of polypharmacy that could contribute to driving impairment" (p. 747). Indeed, Tables 5.1 and 5.2 show the wide variety of prescriptions, supplements, and herbal remedies being used by older adults. The interaction among these medications is unknown in most cases and in need of much more research. One potential way to approach this problem may be to focus on the effects or symptoms produced by the drug doses, combinations, and interactions, rather than on the drugs themselves, and relate the level of these symptoms to declines in critical driving skills. A similar approach has been taken recently in the development of a self-screening instrument for older drivers (Eby, Molnar, & Blatt, 2005; Eby, Molnar, & Kartje, 2007; Eby et al., in press).

Although we have mentioned some names of drugs, we have generally attempted to avoid identifying specific brands. Instead, we refer mainly to drug classes. This is, in part, due to the fact that new drugs are constantly being introduced to the medical field and many older drugs are being discontinued or infrequently prescribed. In large part, the new drugs have not been evaluated on how they affect driving performance or crash risk. It is an inconvenient truth that traffic safety research lags several years behind new drug development. At the same time, drug companies should strive to develop drugs that do not produce drowsiness or sedation as side effects in the older adult population.

A recurrent theme in this chapter, and in this book, is that each individual is unique. Not all medications affect everyone the same way and not everyone has the same resources for managing the effects of medications. Older adults who are taking medications and experiencing drowsiness, sedation, vision problems, movement problems, or thinking problems should discuss them with their physician. They should also discuss their fitness to drive. It is also important to note that physicians should be ready and willing to refer patients to driving professionals, such as certified driving rehabilitation specialists, if they are unsure about giving driving-fitness advice. There is still much progress to be made in the field of medications and driving safety.

6 Screening

[D]river screening is in...theoretical, practical, and ethical terms, a highly dubious activity.

Hakamies-Blomqvist (2006)

INTRODUCTION

One of the most controversial topics in maintaining safe mobility in an aging society is the evaluation of an older person's fitness to drive. The controversy stems from a number of questions: Should evaluation be based on age? Are evaluation instruments valid and reliable? Are the time and financial costs of evaluation worthwhile expenses? Some of the controversy, however, results from a lack of clarity in defining the difference between driver screening and driver assessment. For this book, we use consensus-based definitions developed at the *North American Driver License Policies Workshop* (Molnar & Eby, 2008a):

> Screening and assessment represent different and distinct domains of driver evaluation. Screening is the first step in a multi-tiered process and should not be used to make licensing decisions. Assessment provides the basis for identifying reasons for functional deficits, determining the extent of driving impairment, recommending license actions, and identifying options for driving compensation or remediation (p. 2).

Figure 6.1 presents our conceptualization of the fitness-to-drive evaluation process, including the professionals and venues where screening and assessment take place. Driver evaluation can take place in three venues represented by the three boxes in the figure: home/community, clinical settings, and driver's license agencies. In each of these settings, there are various people who conduct the screening or assessment. The potentially unsafe older driver is generally first identified in his or her community, either through someone observing unsafe driving, an incident such as a crash, or the driver noticing a potential driving problem. In all cases, either formal or informal screening is taking place. At this point the driver may be referred to the licensing agency or may voluntarily seek medical evaluation, participate in an education and rehabilitation program, or self-regulate his or her driving.

This chapter reviews the issues surrounding driver screening. Although much research has been conducted to develop screening tools that are valid and reliable, low cost, and easily administered (see Appendix A), to date, no tool satisfies all of these components. Important research continues on ways to improve the sensitivity (maximizing correct decisions that an individual is a high crash risk) and

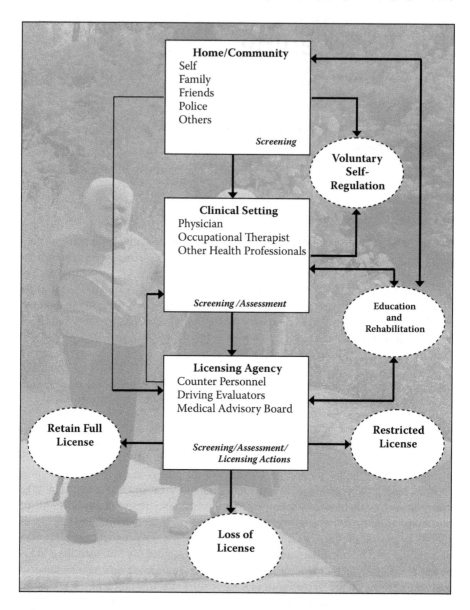

FIGURE 6.1 Schematic representation of the older adult driver screening and assessment process.

the specificity (minimizing incorrect decisions that an individual is a high crash risk) of screening (Staplin, 2008). In the meantime, widespread consensus is that screening tools should focus on detecting functional losses in visual, cognitive, and/or physical abilities that directly affect driving. In addition, screening is increasingly being considered as the first step in a multitiered process for identifying at-risk drivers. In this chapter, we discuss the more promising screening tools that are currently in use.

SELF-SCREENING

There are a number of self-screening tools available designed for use by the older adult driver himself or herself. These tools range from simple paper-and-pencil to sophisticated computerized formats. There are a number of advantages of self-screening (Eby et al., 2003): (1) It is nonintrusive and, therefore, less threatening than other types of screening; (2) because it is nonthreatening, people may be more likely to be screened earlier in disease onset, resulting in earlier detection of functional declines; and (3) self-screening tools, by their very nature, are easily and cheaply distributed resulting in widespread availability. On the other hand, self-screening is not likely to benefit cognitively impaired older adults who lack insight into their abilities, and may even be detrimental in cases in which the results of self-screening are misinterpreted to mean that they are safe to drive when they are not.

TABLE 6.1
List of Self-Screening Tools

Name	Format	Reference/Availability
Driving Decisions Workbook	Booklet	Eby et al. (2003) http://hdl.handle.net/2027.42/1321
Creating Mobility Choices: Older Driver Skill Assessment and Resource Guide	Booklet	AARP (1992) No longer available.
Drivers 55 Plus: Check Your Own Performance	Booklet	AAA Foundation for Traffic Safety (1994). http://www.aaafoundation.org/pdf/driver55.pdf
Safe Driving for Older Adults	Booklet	National Highway Traffic Safety Administration (2003). http://www.nhtsa.gov/people/injury/olddrive/OlderAdultswebsite/index.html
How Is Your Driving Health? A Self-Awareness Checklist & Tips to Help You Drive Safely Longer	Booklet	Staplin et al. (2003a, 2003b) http://www.nhtsa.dot.gov/people/injury/olddrive/modeldriver/1_app_c.htm
The Older and Wiser Driver	Booklet	AAA Foundation for Traffic Safety (1997). http://www.aaafoundation.org/pdf/older&wiser.pdf
Roadwise Review	Computer based	AAA Foundation for Traffic Safety Staplin & Dinh-Zarr (2006) Can be purchased at any AAA office.
SAFER Driving: The Enhanced Driving Decisions Workbook	Web based	Eby et al. (in press) http://www.um-saferdriving.org

Table 6.1 provides a list of self-screening tools, their format, and availability. Only a few of these tools have been formally evaluated. Because these tools are designed to raise self-awareness, we discuss them in more detail in Chapter 9.

FAMILY SCREENING

Family members, friends, and neighbors are often the first to notice or suspect problems with an older driver. Family and friends may gather this information in a number of ways. They may directly observe problems with driving, such as traveling too slow for the speed limit or traffic conditions, stopping mid-block for no apparent reason, or pulling into traffic with too little room. Family or friends may notice new scrapes or dents in the older driver's vehicle, indicating potential driving problems. The older driver may also share information that may indicate there are problems. For instance, an older adult may mention getting annoyed because other drivers are in a hurry and frequently honking their horns at the older driver.

When a family member or friend notices a potential problem, they are, in effect, screening the older driver. At the same time, the family member or friend is placed in the position of deciding what to do about their concerns. There are, fortunately, several guides for helping family and friends address driving with an older adult. As with self-screening, these guides are largely educational and are discussed in more detail in Chapter 9.

SCREENING BY PHYSICIANS

Physicians help their older patients maintain a level of health and fitness by providing effective treatment and preventative health care. This preserves their patients' functional abilities later in life so that their driving years and safe mobility can be prolonged. As the number of older drivers increases, patients and their families will increasingly rely on physicians for guidance regarding safe driving. Physicians will have the challenge of balancing their patients' safety against their needs for mobility and independence, as well as the quality and confidentiality of the physician-patient relationship.

Physicians should take an active role in screening their patients for medical fitness to drive and refer them for a more in-depth assessment if there are concerns. There are difficulties, however, with this process. For many physicians, driving is often overlooked because it is not asked about in the office or hospital setting or it is viewed as being beyond the scope of medical care (Bogner et al., 2004; Marottoli, 2008). Another concern shared among physicians is that if they address driving issues, their patients will not disclose medical problems for fear of losing their driving privileges (Redepenning, 2006; Taylor, Chadwick, & Johnson, 1995). Further, many physicians are aware that screening and assessment tools have not been found to be strongly linked with crash risk and are, therefore, reluctant to use them, while others are unaware of these tools (Marottoli, 2008). Physicians are also concerned that if they screen for driving fitness they will jeopardize their relationships with patients (Molnar et al., 2005). Physicians are particularly concerned about reporting drivers to licensing agencies. A study in Saskatchewan, Canada, found that most physicians would report patients medically unfit to drive, but a majority also believed the patient-physician relationship could be adversely affected by reporting (Marshall & Gilbert, 1999). A focus group study of physicians on assessing older drivers found that physicians are concerned about liability issues if they screen for driving-related issues (Bogner et al., 2004).

Because of these concerns, the American Medical Association (AMA) has published an ethical opinion that expresses the responsibility of physicians to recognize impairments in patients' driving abilities (Wang et al., 2003). The AMA has stated that assessment of physical or mental impairments that might adversely affect driving abilities should happen on a regular basis with the primary care physician when patients are seen for routine physical examinations, as well as when specific medical issues occur. Specialty physicians also have a responsibility for addressing driving fitness with their patients. Ophthalmologists and optometrists regularly assess visual function and should be aware of the visual standards for driving that exist in their state, province, or country so that they can counsel patients whose corrected acuity or peripheral vision do not meet safety guidelines. People who do not pass the visual screening when attempting to renew their license will often be provided a form for their eye specialist to complete and return to the licensing agency. This is one direct opportunity to report visual deficiencies with the patient's permission. Other specialists such as neurologists, emergency room physicians, psychiatrists, physiatrists, and orthopedic surgeons all manage conditions and perform procedures that may have a negative impact on driving ability (see Chapter 4). In addition, all physicians, regardless of specialty, should consider the effects of medications on driving, particularly for their older patients and when prescribed in combination with other medications (see Chapter 5). There are several tools available to help physicians with screening older drivers.

The Physician's Guide to Assessing and Counseling Older Drivers

One screening tool is the *Physician's Guide to Assessing and Counseling Older Drivers,* developed by the AMA and NHTSA (Wang et al., 2003). This guide is designed to provide physicians with information to address the issue of safe mobility in the older patient population. This document presents a flowchart for physician screening, assessment, and remediation titled, "Physician's Plan for Older Drivers' Safety" (PPODS). In this model, screening is the first step in identifying at-risk drivers and involves careful observation of the patient when they present to the physician. The guide recommends that physicians should be alert to poor hygiene and grooming; difficulty with walking or moving from sitting to standing; and difficulty with attention, memory, comprehension, or visual tasks. In addition, physicians are advised to be alert to "red flags" such as any medical condition, medication, or symptom that can impair driving skills either temporarily or permanently. These red flags are further identified as fitting one of the following categories.

- Acute events such as myocardial infarction, stroke, syncope, seizure, or surgery that are generally associated with an emergency room visit or hospitalization

- Patient or family member concerns regarding driving capability; physicians are directed to ask for specific causes of concern as they may indicate a significant safety issue
- Significant medical history due to chronic medical conditions affecting visual, mental, and physical abilities
- Medical history of conditions with unpredictable or episodic events such as hypoglycemia, seizure, or angina
- Medications with strong potential to affect driving performance
- Review of symptoms that may impair driving performance such as fatigue, weakness, tremors, or numbness

If this screening is positive, indicating that the patient is potentially at risk as a driver, the guide recommends that the physician perform a formal assessment called the "Assessment of Driving-Related Skills" (ADReS). ADReS is described in more detail in Chapter 7. A criticism of PPODS is that it was developed based on consensus and best practice rather than on research studies relating recommendations to changes in crash risk.

DETERMINING MEDICAL FITNESS TO OPERATE MOTOR VEHICLES: CMA DRIVER's GUIDE

Several efforts in Canada have resulted in guides and instruments to aid physicians in screening patients for fitness to drive. One such effort is the guide developed by the Canadian Medical Association (2006) called "Determining Medical Fitness to Operate Motor Vehicles: CMA Driver's Guide." At more than 100 pages, the CMA guide provides a wealth of information about medical conditions, medications, alcohol, and driving, as well as advice to physicians on screening and assessment. Despite this detail, the CMA guide has been criticized for providing overly broad recommendations (Hogan, 2005) and for not being evidence based, that is, based on empirical research linking recommendations to decreased crash risk (Molnar et al., 2005).

DRIVING AND DEMENTIA TOOLKIT/SAFE DRIVE

Another effort in Canada by the Dementia Network of Ottawa resulted in the development of the Driving and Dementia Toolkit for primary care physicians (Byszewski et al., 2003). This toolkit consists of background information regarding the older driver and dementia, as well as a list of local resources and how to access them. The toolkit can be found at: http://63.151.41.176/rgap/dementia/task_force_en.asp.

The toolkit includes two sets of screening questions, one for the driver and one for the driver's family member. Also included is a recommended approach to screening called SAFE DRIVE. The title is actually a mnemonic device to encourage thorough screening and stands for:

- **Safety Record:** Obtain a history of driving problems from Department of Motor Vehicles
- **Attention Skills:** Look for lapses of consciousness or recurrent episodes of confusion
- **Family Report:** Ask family members about their observations of patient's driving ability
- **Ethanol:** Screen for alcohol abuse
- **Drugs:** Conduct a medication review, checking for sedating or anticholinergic drugs
- **Reaction Time:** Check for neurologic/musculoskeletal disorders that may slow reactions
- **Intellectual Impairment:** Conduct a Mini-Mental Status Examination
- **Vision/visuospatial function:** Test for visual abilities
- **Executive functions:** Check ability to plan and sequence activities and self-monitor behaviors

An evaluation by the developers of the toolkit (Byszewski et al., 2003) showed that after using the toolkit, physicians' knowledge and confidence regarding dementia and driving significantly increased; physicians were likely to report that they would start following the strategies presented in the toolkit; and physicians were quite satisfied with the toolkit. However, as with other physician tools, the toolkit has been criticized for providing overly broad recommendations (Hogan, 2005) and not being evidence based (Molnar et al., 2005). The screening questions are based on clinical experience and consensus and need to be validated to determine the relationship to driving and traffic safety measures.

CanDRIVE

Another approach, also developed in Canada, is CanDRIVE (Molnar et al., 2005). Similar to SAFE DRIVE, CanDRIVE is also a mnemonic for driver screening. The CanDRIVE acronym reminds that the following areas should be screened (Molnar et al., 2005).

- **C**ognition
- **A**cute or fluctuating illness
- **N**euromuskeletal disease or neurologic effects
- **D**rug use
- **R**ecord (driving)
- **I**n-car experiences
- **V**ision
- **E**thanol use

Molnar et al., (2005) pointed out that like other screening tools, CanDRIVE was not developed based on empirical evidence. They also point out, however, that the Canadian Institutes of Health Research (CIHR) are funding research to develop evidence-based screening tools. When available, this work will be reported at http://www.candrive.ca.

SCREENING BY OTHER HEALTH PROFESSIONALS

Interactions with older adults occur in a variety of different health-care and community living settings. Even health professionals who do not specialize in driving-related issues can play an important part in recognizing impairments and observing the capabilities of older drivers. Social workers, nurses, psychologists, psychiatrists, therapists, and senior housing staff may have the opportunity to help identify potential problems and recommend an older driver screening or assessment. These professionals are also in situations where they can educate the older driver and the family regarding available resources to assist them in maintaining safe mobility. The resources described earlier in this chapter can be utilized by all health professionals. Chapter 9 discusses educational resources to help professionals learn how to work with older adults to assist them in maintaining safe mobility.

SCREENING AT LICENSING AGENCIES

As part of the process they go through with drivers of all ages, licensing agencies have a unique opportunity to regularly screen older drivers during the normal license renewal process. This regularity of screening, however, ranges greatly depending on the jurisdiction (Molnar & Eby, 2005). For example, drivers age 87 or older in Illinois must renew their license yearly, while drivers of any age in Oregon and Wisconsin have to renew only every 8 years. There are also differences among jurisdictions in whether older people need to renew in person versus through the mail. Finally, jurisdictions differ in the type of testing that is required and the standards that are used to determine driving fitness. Chapter 8 specifically addresses issues with the licensing of older drivers. Here we discuss driver screening at licensing agencies, regardless of why the older driver is at the agency. A number of screening programs for use in licensing agencies have been developed and pilot tested.

Model Driver Screening and Evaluation Program

The "Model Driver Screening and Evaluation Program" was developed by NHTSA and the Maryland Department of Motor Vehicles (Staplin et al., 2003a, 2003b; Staplin & Lococo, 2003). This program recommends that screening first be done by licensing agency counterstaff trained to perform a visual inspection of the driver's appearance and functional ability. The inspection should be based on observing the cleanliness and appropriateness of the older drivers' appearance with consideration given to differences driven by socioeconomic backgrounds and culture. The observations should also include ability to move their head, arms, and legs and the ability to walk with or without assistance. The staff member may also ask them questions about their health and medication use and also review their driving history. During this conversation, the staff member further assesses the driver's functional hearing ability, as well as their ability to follow directions and respond to questions appropriately. Throughout the entire interaction, the staff member is observing the behavior of the driver to ascertain that their mental and emotional state makes them a good candidate for continued driving. Depending on the driver's age and state regulations, they may also conduct screening tests for vision, physical, or cognitive deficits that may impair driving.

If gross functional impairments are detected visually, the program recommends that the driver be given an additional functional screening test, complete a driving skills test, and/or file a medical report. Staplin and colleagues (Staplin et al., 2003a, 2003b; Staplin & Lococo, 2003) recommend that the following functional abilities be screened using the screening test listed in parentheses: visual acuity and contrast sensitivity (wall charts, stand-alone testing machine, or computer-based programs); field of view (ophthalmologic perimetry evaluation, useful field of view [UFOV] subtest 2, Scan Chart test); working memory (delayed recall from the Mini Mental State Exam); direct visual search (Trail-Making test part B); divided attention processing speed (Trail-Making B, UFOV subject 2); visualization of missing information (Motor-Free Visual Perception Test); lower limb strength and mobility (Rapid Pace Walk, Foot Tap test); upper body flexibility (Arm Reach test); and head-neck range of motion (Head-Neck Rotation test).

The results of the visual inspection and functional screening by the licensing agency are used to determine the appropriate outcome for the driver. As shown in Figure 6.1, the licensing agency has several options: They can allow the person to renew his or her license with or without provisions, such as a shorter renewal cycle; they can delay the renewal and require documentation from a physician regarding the persons' functional capacity; or they can suspend the license pending a formal driving assessment. Licensing agencies do their best to make decisions that maintain public safety, as well as give consideration to the individual's abilities and circumstances.

CALIFORNIA THREE-TIERED SYSTEM

Another program developed by the California Department of Motor Vehicles is a three-tiered system that incorporates tools to evaluate sensation, perception, and physical and cognitive abilities (Hennessey, 2003). The system has two objectives: (1) to better evaluate fitness to drive for people of all ages; and (2) to avoid taking a driver on a road test if that driver's skills have declined to a dangerous level (Hennessey, 2003). Under this three-tiered system, renewal applicants (and drivers reported to the DMV because of possible problems) are screened for visual acuity, contrast sensitivity, knowledge, and cognitive function, as well as visually assessed lower body impairments and hand/arm impairments. Those who pass the screen are relicensed, while those who fail are referred for medical assessment or remediation. If the physician reports that the driver is fit to drive or their condition benefits enough from remediation, then the driver can be licensed. If the physician reports back that the driver is unfit to drive, the licensing agency will send them to the second tier of evaluation. This tier involves more in-depth testing of cognitive/perceptual abilities and takes longer than the first-tier tests. Those passing the second tier test can be licensed. Those who do not pass can either be referred back to a physician, referred to third-tier testing, or deemed too unsafe to drive. Third-tier testing involves an on-the-road driving evaluation. The results of the pilot test of this program are expected soon.

CONCLUSIONS

Although driver screening has been described as "dubious" (Hakamies-Blomqvist, 2006), it is an inevitable component of maintaining safe mobility in an aging society. At some time, most of us will encounter older parents, relatives, neighbors, or patients who are experiencing declines that may affect driving. We all share in the responsibility of helping our older adult population stay safe and mobile in our communities. As described in this chapter, driver screening is not an exact science. While research continues on the development and validation of effective screening tools, we still have to make decisions in the present based on the best available science and balanced by what is fair and reasonable. At the same time, it is important that we continue to strive for screening tools that have high sensitivity and specificity, and that we consider screening to be the first step in a multitiered process of evaluating fitness to drive for older adults.

7 Assessment

Like everyone else who makes the mistake of getting older, I begin each day with coffee and obituaries.

Bill Cosby

INTRODUCTION

The *American Heritage Dictionary* (2000) provides two definitions for the word *assessment*: "the act of judging a person, situation, or event" and "the classification of someone with respect to their worth." For older adults who are having their driving abilities assessed, many feel as if both definitions are being applied to them. Most older adults have been driving for more than 50 years and have been enjoying the independence and self-reliance that being able to drive affords them. It may never have occurred to these older adults that some day they may not be safe to drive. When screening suggests that their driving may be unsafe they often feel angry and unjustly singled out. For most people, and for older adults in particular, the ability to drive independently is closely related to feelings of self-worth.

As described in Chapter 6 and shown in Figure 6.1, assessment of driving is often triggered by a screening process that identifies potentially unsafe drivers. Assessment is a process that can take place in more than one setting and be administered by more than one person. According to the Association of Driver Rehabilitation Specialists (ADED, 2002), Carr (2000), and others, this assessment generally consists of a clinical evaluation, an on-road evaluation, and postevaluation recommendations. The assessment may be conducted by an occupational therapist (OT), a physician, or other health professional.

CLINICAL ASSESSMENT

The clinical assessment involves establishing a profile of the driver's medical and driving history. The assessment involves an interview and a number of clinical tests (an assessment battery) that evaluate the driver's capabilities in visual, cognitive, and psychomotor functioning.

As discussed by ADED (2002) and Carr (2000), the clinical evaluation serves a number of purposes including establishing the driver's ability to meet the basic criteria for state licensing, determining the strengths and weaknesses of the driver and identifying "red flags" for the road test, identifying the most appropriate type of road test, and determining the driver's potential to benefit from remediation strategies.

THE INTERVIEW

Regardless of who is conducting the clinical assessment, the process should begin with a face-to-face discussion with the older driver, and possibly his or her family member or caregiver. The purposes of the interview are many. First, as discussed by Di Stefano and Macdonald (2006), it is important to understand the goal of the driver evaluation, which can be influenced by individual characteristics and driving-related factors. Second, the medical and driving history of an older adult is an important component of determining his or her fitness to drive. Although medical and licensing agency personnel are in a position to gather objective information about medical history and violation/crash history respectively, others may need to collect this information through an interview. Third, the interview process itself can help in assessing the driver. The way in which a person answers questions, as well as the answers' depth and accuracy, can provide information on the driver's functional memory and insight into any deficits.

The evaluator should keep in mind that the older driver may not always provide accurate or thorough information. Actively engaging a family member or friend, as well as the older adult, may elicit additional input or help correct inaccurate information in a nonthreatening way. Strategically positioning the person's family member as a silent contributor to the interview may be important when the older adult shares limited information or lacks insight. Since the clinical evaluator may see the client only once or twice briefly, this historical information is beneficial for understanding the older adult's needs, abilities, and resources.

According to ADED (2004) and Carr (2000), the interview should

- Determine the older adult's medical history by reviewing past medical records and discussing past medical conditions
- Determine the older adult's current medical status by reviewing current medical records and discussing current/recent medical conditions not included in the medical records
- Determine the client's current use of medications and their side effects by reviewing medical records and discussing medication use with the older adult
- Assess the older adult's communication status by reviewing medical records for hearing-related conditions and talk with the older adult to determine hearing status, ability to comprehend speech, and use of compensatory strategies for communication
- Assess the older adult's driving history by reviewing available driving records and discussing crashes, violations, and typical driving behaviors with the older adult

The occupational therapist may also assess the following during the interview.

- Assess the older adult's license status by inspecting the driver license to confirm that it is valid and contains information on any restrictions or endorsements
- Determine the older adult's driving goals through discussion with the person

- Determine the availability of and types of vehicles available for use by the older adult

In this process, it is particularly important to determine the performance patterns of the older driver. Performance patterns are the habits, routines, and roles adopted by individuals as they engage in occupations (AOTA, 2002a). The habits of a driver are sometimes mentioned in conversation, but are generally easiest to observe during the on-road assessment. Common driving habits include speed control, following distance, and use of directional signals. Understanding older drivers' routines is also important. Asking open-ended questions regarding the frequency of driving, where and how far they travel, the types of roads used, and whether they stay in familiar areas will help the clinical evaluator to understand potential driving challenges.

ASSESSING FUNCTIONAL ABILITIES

According to guidelines published by NHTSA and the American Association of Motor Vehicle Administrators (Staplin & Lococo, 2003) there are three categories of functional abilities that are important for safe driving: vision, cognition, and physical. A study of hundreds of older drivers in Maryland linked assessment outcomes to at-fault crash risk (Staplin, Gish et al., 2003). This study found that the following abilities were important for safe driving: good vision under both high- and low-contrast conditions, good head and neck flexibility, sufficient upper and lower body strength to steer the vehicle and operate the pedals, good working memory, ability to adequately divide attention, ability to visualize missing information, and ability to process visual information quickly.

There is considerable variation in how a clinical evaluator assesses the functional abilities of older adults and in what tools he or she uses. This variability results, in part, from the lack of research identifying the validity, reliability, and specificity of various assessment tools or batteries (Dickerson et al., 2007). The American Medical Association suggests a set of tests that, by consensus, should be used to assess functional abilities needed for safe driving in *The Physician's Guide to Assessing and Counseling Older Drivers* (Wang et al., 2003). The assessment battery is called *ADReS*. This assessment battery includes the following tests: visual acuity; visual fields; rapid pace walk; neck, upper extremity, and ankle range of motion; upper extremity, hip, and ankle strength; Trail Making B test; and the Clock Drawing test. These tests were selected based on their availability, ease of use, and efficiency to administer. The individual tests have been validated as measures for the specific component skills and some have been shown to be related to driving performance.

Work has shown that interrater reliability among various practitioners administering ADReS is high (Posse, McCarthy, & Mann, 2006). Other work addressed the specificity (the probability that the screen is negative given the person does not have a problem) and sensitivity (the probability that the screen is positive given the person does have a problem) of ADReS administered by a sample of physicians (McCarthy & Mann, 2006). This study compared results of ADReS to the outcomes of a behind-the-wheel test with a driver rehabilitation specialist. The study found that ADReS successfully identified all participants who failed the road test but also identified 70 percent of the sample as in need of an intervention. Thus, ADReS classified some people as being problem drivers when they were not. This result is supported by a cross-sectional study of older drivers admitted to an emergency department that evaluated the proportion who had red flags as defined by ADReS (Fender et al., 2007). Of the 50 patients who were assessed, only one had no red flags and 47 could not complete at least one part of the screening battery. The authors concluded that the red flags in ADReS were of limited value, since nearly all patients had them. Until further research on ADReS is conducted, its value as an assessment battery is unclear.

Following the clinical assessment, the evaluator needs to decide whether the older adult should proceed to the on-road portion of the evaluation and what type of road test would be most appropriate. Since there are no standard cut-off scores or objective measures to determine a failure on the clinical test, this decision is usually made based on clinical judgment with consideration of a number of factors including the severity and type of impairments in the component skills, the number of critical skill areas affected, the older adult's self-awareness and insight, and the expected deterioration from chronic illnesses (Hunt, 1996).

ON-ROAD ASSESSMENT

The on-road assessment is considered by many to be the best measure of a person's ability to drive safely, in part because it has high face validity (Kay, Bundy, Clemson, & Jolly, 2008; Stav, Hunt, & Arbesman, 2006). On-road assessments are administered only by trained driving evaluators. The on-road assessment is a contextually based test that allows the evaluator to observe actual driving performance. This performance can be observed over a period of time and in a variety of driving situations. Observing how the driver applies the rules of the road and handles different roadway conditions and challenges is an integral part of the on-road assessment. This assessment might also include determining the need for adaptive equipment (see Chapter 11).

ROUTE SELECTION

On-road assessments can take place in different environments based primarily on the current driving patterns reported by the driver or his or her driving plans for the future including driving only in a familiar area in close proximity to the older adult's home, driving in unfamiliar areas, or driving on a fixed-route course (Di Stefano & Macdonald, 2006). The choice of an appropriate route for on-road assessment is important, as the route's characteristics are what determine how difficult the on-road assessment will be for the driver (Di Stefano & Macdonald, 2006). A road test in a familiar area may be appropriate for the older driver who intends to drive only close to home. This test is sometimes conducted by the evaluator developing a specific local route for the older adult to drive. More often, however, an on-road test in a familiar area is conducted by having the older adult, and possibly a family member, identify typical destinations that the older driver visits. This method is also used frequently for the driver with mild memory impairment who needs to be assessed additionally for memory issues (i.e., remembering where they are going, where they parked the car, etc.) and topographical skills. A road test in an unfamiliar area is intended for a driver who intends to drive wherever he or she wants to go. The fixed-route course is one where the route has been designed to provide a variety of graded traffic situations with different roadways, intersections, and speed zones; it is typically used when evaluating an older driver that intends to drive unrestricted on all different types of roadways. The use of a fixed route allows for more standardized procedures (i.e., directions given) and scoring of the on-road test (Odenheimer et al., 1994).

Di Stefano and Macdonald (2006) also point out that the driver should be given time to adjust to the assessment through the use of "graded routes," that is, routes that become progressively more difficult. The authors suggest that the route should not expose drivers to demanding driving maneuvers and situations for the first 15 minutes of the drive. They also warn that if the driver is not using his or her own vehicle, sufficient practice should be given so that the driver can become familiar with the vehicle controls and handling. Di Stefano and Macdonald (2006) suggest that such practice is best completed in an off-road environment.

ASSESSMENT OF CRITICAL DRIVING SKILLS

As described in Chapter 3, there is a wide range of skills considered critical for safe driving. We categorize these skills as operating on four levels: life goals, strategic, tactical, and operational. Although the last three groups of skills are the ones that can be assessed in an on-road evaluation, the life-goal skills are important to understand when interpreting on-road performance.

The scoring of an on-road assessment is not straightforward. Indeed, there are many variations in how on-road tests are scored (Brown et al., 2005; Di Stefano & Macdonald, 2003, 2006; Fox, Bowden, & Smith, 1998; Hunt et al., 1997; Kay et al., 2008; Lee, Drake, & Cameron, 2002; Schanke & Sundet, 2000). Scoring ranges from ratings such as a global pass/fail, whether or not the evaluator had to intervene, to the use of quantitative scores. Di Stefano and Macdonald (2006) argue that quantitative scores are more reliable and valid but require standardization.

Several researchers have called for greater standardization of on-road driving procedures and scoring (e.g., Di Stefano & Macdonald, 2006; Fox et al., 1998; Kay et al., 2008). Collectively, these researchers recommend the following to improve standardization and, hence, reliability of on-road assessment.

- Use of a standard, predetermined, clearly documented test route
- Scoring of identified aspects of behavior at predetermined points along the route
- Use of standard directions, given in the same form and at the same time, during the test
- Documentation of assessment criteria in specific terms for each required driving maneuver
- Use of a well-defined scoring procedure that has discrete criteria or a quality rating for each maneuver
- Extensive and consistent training for evaluators

Dobbs, Heller, and Schopflocker (1998) and Di Stefano and Macdonald (2003) have emphasized that errors should be evaluated within the context in which they occur and weighted against the errors that healthy, safe drivers make. A study in Australia, for example, investigated the validity and reliability of on-road driving assessment of older drivers (Kay et al., 2008). This study included older adults with specific vision deficits and older adults who were healthy volunteers. The study found that both healthy and vision-impaired older drivers failed to check blind spots before changing lanes. The on-road assessment should focus on those errors that are not made by healthy drivers of a similar age.

POSTEVALUATION RECOMMENDATIONS

A critical step in the driver assessment process is the synthesis of the assessment results, formulation of recommendations, and a postevaluation meeting with the older driver and, if appropriate, his or her family member. As discussed by Carr (2000) the physician has several options for recommendations. The physician can refer the patient to a specialist for further evaluation. For example, if the older driver has suffered a stroke the physician might refer that driver to a neurologist. Depending on the medical condition, the physician can prescribe medication, other treatment options, or rehabilitation to improve driving-related abilities. Carr (2000) also suggests that physicians can refer patients to an occupational therapist who specializes in assessing driving skills, such as a driving rehabilitation specialist, who can also assist older drivers in retraining, vehicle modification, and other community mobility options.

From the perspective of an occupational therapist, the possible outcomes of a driver assessment can fall into several categories that reflect the older drivers' potential to drive independently and safely. The highest performance category is completely independent driving, with no need for intervention or restrictions. In order to be rated at this level, a driver must demonstrate good skills in all performance component areas and the evaluator has no concerns regarding driving safety. The second category is independent with self or formal restrictions. This outcome is for the driver who is able to drive safely under certain conditions. Determination of the restrictions may be based on how the driver already self-regulates his or her driving or they may be based on how the driver did on the on-road assessment. Common restrictions include driving only in familiar areas, staying within a certain geographic radius, restricting nighttime driving, and limiting driving to nonexpressway roads. It should be noted that none of these restrictions were developed based on empirical data showing that the specific restriction improves safety. The third outcome is that the older adult may have the ability to drive safely after some intervention to improve performance. As described by Schold-Davis (2008), there are three types of interventions for driving. The first is remediation, which includes a wide range of possibilities including fitness training, cognitive training, education, medical procedures, and biofeedback. The goal of remediation is to restore functional abilities so that the older adult can again perform critical driving skills adequately. The second type of intervention is compensation, where the driver is taught new strategies to overcome functional deficits. One form of compensation is restricting driving as described previously, but can also include learning better driving habits. The third strategy is the use of adaptive equipment that help to overcome the functional limitation the individual is experiencing. See Chapter 11 for a discussion of vehicle adaptations.

The final outcome is that the older adult cannot independently operate a motor vehicle and remediation, compensation, and adaptation cannot help enough for the person to be able to drive safely. Although maintaining mobility is extremely impor- tant, safety is the overriding factor in making the decision to recommend driving cessation. The evaluator should provide the older adult who has this outcome with information and advice on how to meet mobility needs without driving. See Chapters 13 and 14 for discussions of the issues and options for maintaining safe mobility after driving cessation.

DRIVER ASSESSMENT TEAM

Although the driver assessment process takes many forms across the United States and other countries, a particularly innovative process is to utilize a team approach. The assessment of an older driver is a serious and potentially life-changing issue for older adults. As we have discussed, much of the assessment process outcome is based on clinical judgment and best practice rather than empirical data. Involving a team of professionals allows each member to hear multiple viewpoints in order to make better informed decisions.

The driver assessment team generally consists of a physician who initiates the process by referring the patient for a driving assessment. A written referral or pre- scription is required by most programs and individual practitioners. In this case, the physician consults the driver rehabilitation specialist for an opinion and recom- mendation. The clinical and on-road assessments are usually completed by a driver rehabilitation specialist. This is an occupational therapist that has additional levels of knowledge and expertise beyond the normal training of an occupational thera- pist (OT). The on-road assessment or training, if recommended, may also involve a driving instructor. Some program models utilize a driving instructor in the front seat during the on-road test while the driver rehabilitation specialist observes and scores the assessment in the back seat. Other effective models of practice include the same driver rehabilitation specialist completing both assessments and two different driver rehabilitation specialists—one to complete the clinical and the other to do the on-road assessment. If it is determined that the driver requires adaptive equipment to physically manage vehicle controls, then both a mobility equipment dealer and the licensing agency should become part of the team. The equipment dealer will be involved in installing and maintaining the necessary driving equipment while the OT would provide any driver training. Each state's licensing agency has procedures for licensing drivers with equipment. The licensing agency may also get involved when either driving restrictions or cessation are recommended.

CONCLUSIONS

The assessment of a driver's fitness to drive is perhaps one of the most important aspects of maintaining safe mobility in an aging society. Yet it is also one of the most subjective. In the absence of objective empirical data on which to base decisions about driving fitness, evaluators are forced to rely heavily on clinical judgment. Fortunately, driver rehabilitation specialists receive specialized training in order to make these judgments. At the same time, the lack of standardization among evaluators, both in the administration and scoring, of the driver assessment makes it difficult to determine the validity and reliability of assessment practices. There is a great need to identify best practices in the field of driver assessment and to conduct evaluative research to determine if the practices are accurately identifying at-risk drivers while also allowing drivers who are safe to continue driving.

8 Licensing, Legal, and Policy Issues

Baseball is like driving; it's the one who gets home safely that counts.

Tommy Lasorda

INTRODUCTION

One primary role of licensing agencies is to make sure that drivers are capable and competent to operate a motor vehicle (Snook, 2008). This stems from the generally accepted view in American society that driving is a privilege and therefore those who cannot drive at an acceptable level of safety should not be allowed to drive (Eby & Molnar, 2008a). However, the public safety role of licensing agencies must be carried out in an environment characterized by constraints on time, budgets, staff availability, staff expertise and training, hardware, and real estate (Staplin, 2008), and often competes with other priorities such as serving customers efficiently with regard to vehicle registration and other nonsafety licensing issues.

Licensing agencies are also being called on to prepare for the increasing numbers of older adults who will be driving and may eventually lose their license because of functional declines (Carr, 2008). As discussed elsewhere in this book, serious adverse consequences can arise from having to give up driving including decreased socialization or a reduction in out-of-home activities (Marottoli, Mendes de Leon, Glass, Williams, Cooney, & Berkman, 2000), increased depressive symptoms (Ragland, Satariano, & MacLeod, 2005), increased risk of nursing home placement (Freeman, Gange, Muñoz, & West, 2006), and increased societal costs for providing transportation services to older adults who no longer drive. Because of these consequences, there are clear personal and societal benefits in expanding the role of licensing agencies beyond simply identifying at-risk drivers to include helping them maintain safe driving for as long as possible and assisting them in transitioning to nondriving when they are no longer able to drive safely (Carr, 2008; Silverstein, 2008).

Licensing agencies have a unique opportunity to identify at-risk drivers because older drivers, like everyone else in the driving population, must go through a license renewal process. Licensing agencies also rely on review of driver history records and referrals from health professionals (e.g., physicians, occupational and physical therapists, social workers, vision specialists), law enforcement officers, courts, and families and friends of older drivers to alert them to situations in which an individual's driving fitness may be in question. Based on the outcomes of these various activities, licensing agencies have several choices: They can allow a driver to keep his or her license; refuse to renew the license or suspend, revoke, or restrict the license (e.g., prohibit night driving, require vehicle adaptive equipment, restrict driving to specific times or distances from home); or shorten the renewal cycle. In making these choices, licensing agencies may consider each individual's abilities and circumstances, and the options available for driving compensation and remediation, as well as rely on the advice of their state medical advisory board if one is in place.

This chapter reviews two important licensing policies related to the identification of at-risk drivers: driver's license renewal and physician reporting of at-risk drivers. It also highlights approaches to strengthening licensing policy and concludes with a discussion of how the expanded role of licensing agencies can be enhanced by a collaborative approach between the licensing agency and a whole host of individuals and groups within communities with an interest in the safe mobility of older adults.

DRIVER'S LICENSE RENEWAL

Driver's license renewal policies in the United States vary from state to state in terms of the length of the renewal cycle, requirements for accelerated renewal for older drivers relative to other drivers, and other renewal provisions (see Table 8.1). Most states and the District of Columbia require that drivers renew their license every 4 or 5 years (19 and 18 jurisdictions, respectively), although seven states have up to a 6-year renewal, four states up to an 8-year renewal, two states a 10-year renewal, and one state (Arizona) no renewal until age 65. Fourteen states require accelerated renewal for older drivers. The beginning age for accelerated renewal ranges from 61 (Colorado) to 81 (Illinois), and the length of the accelerated renewal cycle ranges from 1 year (Illinois for age 87 and older) to 5 years (Arizona, Colorado, South Carolina). One state (Tennessee) actually has decelerated renewal for older drivers with no renewal required after age 65.

Seventeen states have other special renewal provisions for older drivers, including requirements for in-person renewal, vision tests, and other testing or certification (e.g., written and road tests, certification of fitness). In many states, the vision test requirement can be met either through testing in the licensing agency itself or by providing results of a test performed elsewhere. The beginning age for renewing in person ranges from 65 (Connecticut) to 70 (Arizona, California, Louisiana). The age at which a vision test is required at renewal ranges from 40 (Maryland) to 80 (Virginia). Illinois and New Hampshire require a road test at renewal for drivers age 75 and older.

Given that driver's license renewal provisions represent public policies aimed at increasing the driving safety of the public, it is important to ask how well these policies are doing in reducing crash-related injuries and deaths, particularly among older drivers. Unfortunately, research on the effects of older driver license renewal provisions has been limited. There is some evidence that vision testing at renewal may be associated with reduced motor vehicle deaths among older drivers (e.g., see Levy, Vernick, & Howard, 1995). However, a recent examination of renewal provisions in the contiguous United States found that only in-person renewal was related to reduced fatalities, and only among the oldest-old (age 85 and older; Grabowski, Campbell, & Morrisey, 2004). Vision tests, road tests, and accelerated license renewal did not

TABLE 8.1
Licensing Provisions by State for Older Drivers

State	Length of Renewal Cycle (Years)	Accelerated Renewal for Older Drivers (Years)	Other Renewal Provisions
Alabama	4	No	None
Alaska	5	No	No mail renewal for age 69 and older; no more than one mail renewal in a row for all ages; vision test required at renewal for all ages
Arizona	Until age 65	5 for age 65 and older	No mail renewal for age 70 and older; vision test verification required for age 65 and older mail renewal; vision test required every 12 years for all ages
Arkansas	4	No	Vision test required at renewal for all ages
California	5	No	No mail renewal for age 70 and older; no more than two successive mail renewals for all ages
Colorado	10	5 for age 61 and older	No mail renewal for age 66 and older or electronic renewal for age 60 and older; no more than one mail/electronic renewal in a row for all ages
Connecticut	4 or 6	Age 65 and older may choose 2 or 6 yr.	Mail renewal for age 65 and older only if show hardship; vision test required at first renewal and then every other renewal for all ages
Delaware	5	No	None
Dist of Columbia	5	No	Physician certification of physical/mental driving competency, vision test, and possible reaction test required at renewal for age 70 and older; written and road tests may be required at renewal for age 75 and older
Florida	6, 4—bad record	No	Vision test required at renewal for age 80 and older; no more than two successive mail/electronic renewals for all ages
Georgia	4	No	Vision test required at renewal for all ages; mail/electronic renewal every other renewal for all ages
Hawaii	6	2 for age 72 and older	Vision test required at renewal for all ages

TABLE 8.1 (continued)
Licensing Provisions by State for Older Drivers

State	Length of Renewal Cycle (Years)	Accelerated Renewal for Older Drivers (Years)	Other Renewal Provisions
Idaho	4	4 or 8 for age 21–62; 4 for 63 and older	Vision test required at renewal for all ages
Illinois	4	2 for age 81–86; 1 for age 87 and older	Road test required at renewal for age 75 and older; vision test required for in-person renewal
Indiana	4	3 for age 75 and older	Vision test required at renewal for all ages; electronic renewal every other renewal if meet eligibility criteria
Iowa	5	2 for age 70 and older	Vision test required at renewal for all ages
Kansas	6	4 for age 65 and older	Vision test required at renewal for all ages
Kentucky	4	No	None
Louisiana	4	No	No mail renewal for age 70 and older; no more than one mail renewal in a row for all ages
Maine	6	4 for age 65 and older	Vision test required at every other renewal for age 40–61 and at every renewal for age 62 and older
Maryland	5	No	Vision test required at every renewal for age 40 and older; age 70 and older new licensees must show proof of prior safe car operation or physician's certification of fitness; age alone not grounds for re-examination
Massachusetts	5	No	Age discrimination w/ regard to licensing prohibited
Michigan	4	No	Vision test required at in-person renewal for all ages; no more than one mail renewal in a row for all ages
Minnesota	4	No	Vision test required at renewal for all ages; age alone not grounds for re-examination
Mississippi	4	No	None
Missouri	6	3 for age 70 and older	Vision test required at renewal for all ages

TABLE 8.1 (continued)
Licensing Provisions by State for Older Drivers

State	Length of Renewal Cycle (Years)	Accelerated Renewal for Older Drivers (Years)	Other Renewal Provisions
Montana	8 (4 by mail)	4 for age 75 and older	Vision test required at renewal for all ages; mail renewal for all ages only in areas with no driver license services—no more than one in a row
Nebraska	5	No	Vision test required at renewal for all ages
Nevada	4	No	Medical report required at mail renewal for age 70 and older; no more than two successive mail/electronic renewals for all ages; age alone not grounds for re-examination
New Hampshire	5	No	Road test required at renewal for age 75 and older
New Jersey	4	No	Vision test may be required at renewal for all ages
New Mexico	4 or 8	4 if turn 75 in 2nd half of 8-year renewal cycle	Vision test may be required at renewal for all ages
New York	5	No	Vision test required at renewal for all ages
North Carolina	5	No	Parallel parking not required in road test for age 60 and older; vision test required at renewal for all ages
North Dakota	4	No	Certification of vision required at renewal for all ages
Ohio	4	No	Vision test required at renewal for all ages
Oklahoma	4	No	License fees reduced for age 62–64, waived for age 65 and older
Oregon	8	No	Vision screening required every 8 years for age 50 and older
Pennsylvania	4	Age 65 and older may choose 4 or 2 yr.	Vision test may be required at renewal for all ages
Rhode Island	5	2 for age 70 and older	None

TABLE 8.1 (continued)
Licensing Provisions by State for Older Drivers

South Carolina	10	5 for age 65 and older	Vision test required at renewal for age 65 and older; beginning Oct. 1, 2008, vision test required every 5 years for all ages
South Dakota	5	No	Vision test required at renewal for all ages
Tennessee	5	No	No expiration for licenses issued to age 65 and older; no more than one mail/electronic renewal in a row at all ages; fees reduced age 60 and older
Texas	6	No	Vision test required at renewal for all ages
Utah	5	No	Vision test required for age 65 and older; vision test required every 10 years for all ages; no more than one electronic renewal in a row for all ages
Vermont	4	No	None
Virginia	5	No	Vision test required at renewal for age 80 and older; no more than one mail/electronic renewal in a row for all ages
Washington	5	No	Vision test required at renewal for all ages; no more than one mail/electronic renewal in a row for all ages
West Virginia	5	No	None
Wisconsin	8	No	Vision test required at renewal for all ages
Wyoming	4	No	Vision test required at renewal for all ages; no more than one mail renewal in a row for all ages

Adapted from Molnar, Eby, St. Louis, and Neumeyer (2007).
Sources: American Automobile Association, 2005; Insurance Institute for Highway Safety, 2004; and individual state driver licensing Web sites.

result in additional benefits. A recent review by Langford and Koppel (2006b) of several studies in Australia and elsewhere (including Grabowski et al., 2004; Hakamies-Blomqvist, Johansson, & Lundberg, 1996; Langford, Fitzharris, Koppel et al., 2004 et al., Torpey, 1986) also failed to find safety benefits associated with mandatory testing of older drivers.

The lack of supportive evidence for the effectiveness of specific renewal provisions, coupled with the cost constraints faced by licensing agencies, may help to explain the variety of policies in place around the country. Testing can be expensive to implement and when clear guidelines are lacking about the best way to identify at-risk drivers during the renewal process, there may be little incentive for states to re-examine their existing policies. Further research on how to improve the safety benefits of license renewal provisions for older drivers is warranted.

PHYSICIAN REPORTING OF AT-RISK DRIVERS

The issue of physician reporting has both ethical and legal implications. From an ethical standpoint, the American Medical Association recommends that physicians notify their state licensing agency when patients fail to appropriately self-regulate their driving or when they otherwise perceive a threat to public safety; however, physicians must be able to identify and document physical or mental impairments that clearly relate to the ability to drive (Staplin, 2008). From a legal standpoint, few states require physicians or other professionals to report at-risk drivers to licensing agencies (Carr, 2008). Instead, most states rely on voluntary procedures to identify at-risk drivers that involve the reporting of concerns about individual drivers to licensing agencies by professionals of various types or family members (Meuser, Carr, Berg-Weger, Niewoehner, & Morris, 2006; Morrisey & Grabowski, 2005). A small number of states, including California, Delaware, New Jersey, Oregon, and Pennsylvania, require physicians to report certain health conditions (e.g., Alzheimer's disease, epilepsy) to the licensing agency at the time of diagnosis so that ongoing driving fitness can be addressed (Wang et al., 2003). Licensed drivers themselves are also encouraged to self-report any concerns about their driving to the licensing agency and/or self-limit their on-road exposure (Meuser, 2008).

The specific provisions of voluntary reporting laws vary across jurisdictions. Meuser (2008) recently reviewed this literature. He noted, for example, that Missouri's law allows health professionals, law enforcement officers, family members, and others to report potentially unsafe drivers for medical evaluation and testing. The law provides confidentiality for the reporter (i.e., the reporter's identity is held in confidence from the reported driver) and civil immunity from prosecution for breach of confidentiality. A similar law in New York allows for reporting by these groups, but does not keep the reporter's identity confidential, nor does it offer civil immunity protection. The author pointed out that little is known about how such provisions impact both physician behavior and public safety. The first comprehensive evaluation of state voluntary reporting laws, linking individual crash and citation histories with medical and behavioral data from over 5,000 drivers reported as unfit in Missouri from 1999 to 2005, is under way (Meuser et al., 2007) and is intended to inform development of a model voluntary reporting process for all states.

A recent paper by Staplin (2008) also highlighted the lack of uniformity with respect to physician reporting policies and practices. He noted that, in some states, the only guidance about disqualifying conditions (apart from vision impairments) that licensing agencies provided to physicians related to seizure disorders/loss of consciousness; other jurisdictions (Utah, Maine, North Carolina) offered contrasting approaches addressing multiple medical conditions, based on "functional ability pro-files" (Lococo & Staplin, 2005). Close to half the states provide some type of protection from civil liability for physicians, while a fewer number offer legal protection or confidentiality (Molnar & Eby, 2005). However, if a patient is involved in a crash and determined to be at fault, physicians can be held liable—even in states with voluntary reporting; the key for physicians to protect themselves from liability is to document all concerns, recommendations, and referrals to outside sources and keep them in the patient's file (Carr, 2008). Many physicians are not aware of reporting requirements in their state. A recently developed resource for physicians and other professionals is the *Physician's Guide to Assessing and Counseling the Older Driver* from the American Medical Association (discussed elsewhere in this book), which details the reporting regulations and mechanisms for each state (Wang et al., 2003).

APPROACHES TO STRENGTHENING LICENSING POLICY

A recent workshop on licensing policy sponsored by the AAA Foundation for Traffic Safety led to a number of consensus-based recommendations, based on the best available science. The workshop was intended to build on several recent complementary events focusing on licensing policies for older adults, including "Challenging Myths and Opening Minds Forum: Aging and the Medically At-Risk Driver," a workshop sponsored by the American Association of Motor Vehicle Administrators (AAMVA, 2006); "Driving and Function Forum," a workshop sponsored by the Office of the Superintendent of Motor Vehicles (OSMV) and the Traffic Injury Research Foundation (TIRF, 2006); and "Licensing Authorities' Options for Managing Older Driver Safety—Practical Advice from the Researchers," a workshop that took place at the Transportation Research Board (TRB) Annual Meeting (Langford, Braitman, Charlton, Eberhard, O'Neill, Staplin, & Stutts, 2007[1]); as well as published literature including "Strategies for Medical Advisory Boards and Licensing Review" (Lococo & Staplin, 2005).

Consensus-based recommendations for licensing policy resulting from the AAA Foundation for Traffic Safety workshop included the following (Eby & Molnar, 2008):

- Base final licensing decisions on functional and medical fitness to drive (and not chronological age)
- Develop through additional research and implement across jurisdictions empirically defensible criteria and guidelines on medical and functional fitness to drive
- Enact standard reporting laws that provide civil immunity for clinicians and licensing personnel who report people they think may be medically unfit to drive; such laws would remove one barrier to reporting—fear of lawsuits
- Establish and fund active medical advisory boards, which should be an integral element of state licensing agencies and should be involved in both case review and policy development

---

[1] Several articles from this workshop appear in *Traffic Injury Prevention*, 9(4).

- Expand the role of licensing agencies to include assisting at-risk drivers to transition from driving themselves to using other community mobility options

Workshop participants also identified recommended elements of model licensing systems. Top choices included the following.

- Driver assessment should not be age determined, but triggered by decreasing functional ability, as measured objectively through screening
- Safety (crash prevention) should serve as the primary basis for driver screening and assessment
- Although it is not appropriate (or practical) to have age-triggered assessment, it is appropriate to have age-triggered driver screening, with screening used only to see if further testing should be done, not to determine license actions that can have much wider ramifications
- In-person driver's license renewal should be required for drivers of all ages
- A medical advisory board with broad representation should be involved in both the evaluation of driver performance and policy development relative to licensing
- Voluntary reporting of at-risk drivers to licensing authorities is important, as is immunity for those reporting

The consensus-based recommendations that came out of the workshop were largely consistent with outcomes from the other licensing policy-related events mentioned earlier in this chapter. Findings from the 2007 TRB workshop highlighted the promise of multitiered systems supported by active medical advisory boards, and suggested that driver screening may be useful as part of a comprehensive medical review but should not be relied on for categorical licensing decisions. Licensing agencies were advised to consider becoming more involved in helping at-risk drivers transition from driving to using alternative transportation options, possibly in conjunction with other agencies. Other advice included basing driver licensing policies and practices on timely empirical evidence, and in the case of conflicting results, taking into consideration the mobility, independence, and social needs of older adults, as well as public perceptions of fair play, in licensing decisions.

The 2006 Canadian Driving and Function Forum also highlighted the importance of multitiered systems, as well as the need for a continuum of assessment rather than just pass/fail and multiple settings for identifying those who may be impaired (e.g., licensing agencies, health-care settings, public health agencies, families, community). Key issues focused on the need to distinguish between screening and assessment; consider the setting and infrastructure requirements; establish validity and reliability, sensitivity, specificity, and predictive power; standardize administration within and across settings; and conduct structured observations during the license renewal process. Further research was recommended to address limitations such as the need for more evidence-based tools and the validation of existing tools.

Finally, the 2006 AAMVA forum led to identification of a set of underlying core values to guide licensing policy for screening and assessment that included safety, basis on fitness not age, sustainability, practicality, flexibility, basis on evidence, acceptability to the public, and routine re-evaluations. Among the conclusions from the forum were the following: Age-based criteria are an inefficient and costly screening mechanism; fitness-based criteria will address the problems present among older drivers; medical conditions that affect driver fitness can occur at any age; driver cessation programs are necessary; wellness programs can extend driving; and everyone will stop driving one day, which underscores the need to plan ahead.

LICENSING AGENCIES AS PART OF A SYSTEMS APPROACH TO SAFE MOBILITY

While licensing agencies have the ultimate responsibility for determining the fitness to drive of older adults, it is clear that they need to work hand in hand with a larger system of individuals and groups within the community. In this regard, Silverstein (2008) has called for a paradigm shift—with the scope of responsibility shared between the healthcare system, licensing authorities, and the community. In her view, the identification of at-risk drivers must be a collaborative effort between citizens, professionals, and government officials, although the final decision to revoke an individual's license to drive resides solely with the licensing agency. Staplin (2008) similarly noted the increasing recognition that the effectiveness of the transportation system requires a *system* solution—namely, an evolution and harmonization of licensing policies, coupled with more "human-centered" principles of highway design and operations (see also Staplin & Freund, 2005). Australia has made considerable strides in promoting a safe systems approach to keeping older drivers safely mobile. As outlined by Langford and Oxley (2006), Australia's safe system framework consists of (1) safer roads, through a series of design improvements particularly governing urban intersections; (2) safer vehicles, through both the promotion of crashworthiness as a critical consideration when purchasing a vehicle and the wide use of developed and developing ITS technologies; (3) safer speeds especially at intersections; and (4) safer road users, through both improved assessment procedures to identify older drivers with reduced fitness to drive and educational efforts to encourage safer driving habits, particularly but not only through self-regulation of driving.

Related to this approach is the view that fitness-to-drive policy must be considered within the broader framework of community mobility. For example, Classen and Awadzi (2008) have recommended that licensing policy for older drivers, as a way to address and promote safe driving, must be viewed in the broader spectrum of independent community mobility. Thus, if policies are going to target older adults who can no longer drive, solutions such as counseling and provision of alternative transportation options must go hand in hand with such policies.

CONCLUSIONS

Licensing agencies have the responsibility to ensure that individuals who are licensed to drive can do so at an acceptable level of safety. With regard to older drivers, licensing agencies carry out this responsibility through various policies and practices including driver's license renewal and physician reporting of at-risk drivers. These policies and practices vary considerably across jurisdictions, and a widely accepted model system for licensing has yet to emerge. However, there is growing consensus about where licensing policy should be headed, including the need for empirically defensible guidelines on fitness to drive; standard reporting laws (with voluntary reporting of at-risk drivers by physicians/clinicians, other professionals, families, and older drivers themselves); and active medical advisory boards to assist licensing agencies in case review and policy development. There is also an opportunity for licensing agencies to expand their role in assisting older drivers' transition from independent driving to other community mobility options. Further research is needed to evaluate promising approaches for helping licensing agencies identify and deal with at-risk drivers, including the capacity and willingness of various groups to report at-risk drivers, as well as the effectiveness of restricted licensing in keeping older drivers safely mobile.

9 Education and Rehabilitation

Anyone who stops learning is old, whether at twenty or eighty. Anyone who keeps learning stays young.

Henry Ford

INTRODUCTION

The aging process affects everyone in one way or another and many older adults will eventually be faced with questions about their ability to continue to drive safely. How they answer these questions, and whether they are even willing to consider them, depends to a great extent on the information available to them about functional declines in abilities that can affect driving; strategies for compensating for, or overcoming, these declines; and how to plan for a time when driving is no longer possible. Older adults with dementia or other cognitive impairments, however, may lack the insight to benefit from these educational efforts. In addition, as discussed in Part 3 of this book, the older adult's family members and the medical professional also play a role in helping to maintain safe mobility for aging drivers. The effectiveness of these groups' ability to help older drivers is dependent upon the quality of information and training they receive. Thus, the availability of sound education and rehabilitation is essential for maintaining mobility among older adults.

EDUCATION

The English philosopher Sir Frances Bacon (1597) wrote the famous Latin phrase *scientia potentia est*, or "for also knowledge itself is power." Usually he is misquoted as "knowledge is power." For society today, however, knowledge is also a key to maintaining safe mobility throughout the lifespan. Here we review the available education and training programs for the older driver, the driver's family and friends, and health professionals. The literature inconsistently distinguishes between education (the acquisition of knowledge) and training (hands-on skill acquisition), yet we know there are clear differences, particularly in terms of the impacts on behavioral changes. Here we will discuss the issues surrounding education for the driver him or herself, the driver's family and friends, and health professionals.

FOR THE DRIVER

There are several educational programs available for the older driver. Educational resources vary widely in terms of purpose, format, and content. While many older drivers recognize their declining abilities and take steps to adjust their driving, others are unaware of the changes they are experiencing and the implications of these changes for safe driving. Thus, one focus of many education programs is simply to increase older drivers' awareness and knowledge about these issues. Other programs combine education with some type of training to help older drivers compensate for, or when possible, overcome functional declines. Such education and training programs are often part of the rehabilitation process for older drivers identified as functionally impaired by occupational therapists, certified driving rehabilitation specialists, and other professionals.

Driver refresher courses are also available. These programs use classroom instruction to reinforce older drivers' existing driving skills and knowledge, and teach them about newer traffic laws and practices for defensive driving. National programs of this type include the Driver Safety Program course, sponsored by AARP, Safe Driving for Mature Operators, sponsored by AAA, and Coaching Mature Drivers, sponsored by the National Safety Council. On-road driver training programs for older drivers focus on enhancing driving skills by providing opportunities for behind-the-wheel practice. Programs of this type include the Driving School Association of the Americas, the Driver Skill Enhancement Program, and AAA driving refresher courses.

Molnar et al. (2007) have defined criteria they consider important for a promising approach for education programs.

- Program development/design based on age-related research
- Strong marketing approach to ensure public awareness of program and participation by older drivers
- Accessibility to older people through a broad array of settings
- Incorporation of basic learning principles into program design

Effective education programs should build on what is known about functional declines, how they affect driving, and what can realistically be done to address the declines. In the case of declines that cannot be reversed, this may mean simply increasing knowledge and personal awareness so that older drivers can make informed decisions about how to recognize declines, and how to reduce or stop their driving if safe driving is no longer possible. When declines can be overcome, it may mean teaching older people to do new things or to do things in a different way (e.g., learning to do stretching and strengthening exercises). Sometimes the focus of education is on general driving skills that need to be improved because of a lack of knowledge about newer traffic laws or roadway design changes. In these scenarios, the goal of the program is to provide the necessary information and practice opportunities to improve driving skills.

Regardless of the program's focus, older drivers must not only be made aware of the program, but also believe that they can benefit from it. Many older drivers are unwilling or unable to recognize deficits in driving-related abilities, but self-screening is often a useful first step in getting people to take action about their driving. The challenge is to get people to participate, which requires programs that are effectively marketed. One successful approach for doing this has been to apply the principles of consumer marketing to the promotion of health and safety behavior, a practice called *social marketing*. Components of a social marketing approach include situational analysis (getting to know the market), market segmentation and targeting, setting objectives, formulating a market strategy, program implementation, and monitoring/evaluation (Milton, 2004).

Programs must also be accessible to the older adult population they are trying to serve. One way of making programs accessible is to offer them through existing programs or organizations that are already known and used by the older adult. Fitness training programs, for example, can be developed and delivered through existing community or senior centers, recreation centers, public health departments, housing authorities, and religious institutions.

It is also critical that education and training efforts take into account what is known about how people learn, especially older adults. The University of Louisiana at Monroe School of Nursing (2008) provides some excellent advice and guidelines.

- Learning may be slower for the older adult—but it can be accomplished
- The older adult is motivated to learn if it will allow him/her to remain independent
- The older adult brings a lifetime of problem-solving skills to the learning situation
- The older adult may have visual and hearing difficulties that may interfere with learning
- The older adult generally performs tasks slower
- The older adult generally shows great care and concentration, and may have to sacrifice speed for accuracy to minimize the risk of error
- Instructions or information should be given in small units
- Demonstrations should be used for explanations and to facilitate comprehension

Molnar et al. (2007) conducted an extensive review of formal educational programs. Shown in Table 9.1 are some of the more widely available programs considered promising.

Are formal educational programs for older drivers effective? Most educational programs gather feedback from participants about how much they liked the programs and ways to improve it. These data are useful for program developers but do not provide objective data for program effectiveness. They fail to address the core outcomes of programs: amount learned, changes in behaviors due to the program, and changes in motor vehicle crash/injury risk. These outcomes should be addressed in a formal evaluation. Of those that have been formally evaluated, the research shows that educational programs

- Increase the driver's knowledge and awareness (Eby et al., 2003; Eby et al., in press; Marottoli, 2007; Marottoli, Van Ness et al., 2007; Owsley et al., 2003; Stalvey & Owsley, 2003)

TABLE 9.1
Driver Training and Education Programs

Program Name	Organization and Web Site	Program Objective
Driver Safety Program	AARP *www.aarp.org/drive*	To help drivers enhance existing skills and develop defensive driving techniques
Mature Driver Improvement Program	California Department of Motor Vehicles www.dmv.ca.gov/vehindustry/ol/mature_drvr.htm	To educate drivers on the effects of medication, fatigue, alcohol, and visual/auditory impairments on driving performance, and provide strategies for defensive driving
Mature Driver Workshop	Traffic Improvement Association www.tiami.org/maturedriver.asp	To help older drivers evaluate and improve their driving skills, and continue to drive for as long as they can safely do so
Coaching the Mature Driver	National Safety Council www.nsc.org	To educate drivers about ways to adapt to technological changes in vehicles as well as about physical changes commonly associated with aging
Safe Driving for Mature Operators	AAA Traffic Safety Programs www.AAA.com	To meet the informational needs and interests of drivers over age 55 and help extend their safe driving careers
Super Seniors	Illinois Secretary of State www.sos.state.il.us/home.html	To connect training and licensing so that older adults can continue to drive safely longer
Years Ahead—Road Safety for Seniors Program	Royal Automobile Club of Queensland www.racv.com.au	To promote behaviors conducive to safe driving, encourage older drivers to take responsibility for the future of their driving, increase knowledge and awareness, and promote self-screening of driving abilities
Wiser Driver Program	Hawthorn Community Information Centre www.hcec.edu.au	To educate and provide assistance to older drivers regarding road safety and provide information about alternatives to driving
DriveWell	American Society on Aging www.asaging.org/drivewell	To promote older driver safety and community mobility
GrandDriver	American Association of Motor Vehicle Administrators www.granddriver.info/	To help prepare drivers to maintain safe mobility in their later years
Project Safe R.O.A.D.s	Onondaga County Department for Aging and Youth www.ongov.net/Aging_and_Youth/SafeRoads	To provide information that will help keep older drivers safe and to provide access to information on transportation options

TABLE 9.1 (continued)
Driver Training and Education Programs

Program Name	Organization and Web Site	Program Objective
Road Map to Driving Wellness	American Society on Aging www.asaging.org/CDC/ module4/home.cfm	To provide a balanced picture of the functional challenges that face many older adults and an array of ways in which older adults can take action to enhance their driving safety and remain mobile and connected to their communities

- Increase safe driving behaviors, by self-report (McCoy et al., 1993; Owsley et al., 2004; Owsley et al., 2003)
- Improve on-road evaluation scores (Bédard et al., 2004; Marottoli, 2007; Marottoli, Van Ness, et al., 2007)
- Do *not* help to prevent roadway injuries or crashes (Berube, 1995; Ker et al., 2005; Kua et al., 2007; Nasvadi & Vavrik, 2007; Owsley et al., 2004)
- May *increase* the number of crashes for men age 75 and older (Nasvadi & Vavrik, 2007)

The latter two findings may be surprising to some readers. There are, however, at least two possibilities why educational programs do not seem to lower crash risk and may actually increase it. First, because crashes are rare events, to find differences in crash rates, studies require many program participants with many years of driving history data. Such studies are prohibitively expensive for many evaluators. Second, one difficulty in studying any program is selection bias (Ker et al., 2005; Nasvadi & Vavrik, 2007). Educational programs are voluntary and those who elect to participate may be worse (or better) drivers than the general driving population. In other words, if all participants are poorer drivers than the rest of the population to begin with and the program improves driving safety up to the level of everyone else (a safety benefit), then there should be no difference in crash rates. However, the Nasvadi and Vavrik (2007) study controlled for self-selection bias and still found increased crash rates for drivers age 75 and older. Without further research we can only speculate why this occurs. Perhaps educational programs may increase confidence in this age group, exposing them to greater risk of crash (Hunt & Arbesman, 2008). It is premature to give up on these types of programs, so more research is needed in this area.

Driver Self-Awareness Tools

Another type of educational tool for older drivers is the class of instruments that allow older drivers, in their own home, to learn more about functional declines they may be experiencing and what they can do about them. As discussed in Chapter 6, there are several tools available for self-screening. These tools also serve as a way for drivers to increase self-awareness. As described by Eby and Molnar (2005), there are several advantages to self-screening tools: They can be completed in private; people may be more likely to complete the tool earlier in disease onset, resulting in earlier detection of functional declines; and self-awareness tools are easily and cheaply distributed, resulting in wide availability. The main disadvantage is that cognitively impaired individuals lack the insight to benefit from them and they might misinterpret the information and conclude that they are safe to drive when they are not.

One of the promising approaches to self-awareness education is the *Driving Decisions Workbook*, a paper-and-pencil, self-screening instrument developed by the University of Michigan Transportation Research Institute (UMTRI; Eby et al., 2000, 2003). The workbook is designed to increase older drivers' self-awareness and general knowledge about driving-related declines in abilities, and to make recommendations about driving compensation and remediation strategies that could extend safe driving, as well as further evaluation that might be needed. Development of the self-awareness instrument was based on review of the literature, focus groups with older drivers and the adult children of older drivers, and a panel of experts on older driver abilities and evaluation. Based on findings from these activities, a model of the influences on driving decisions was developed with three domains for screening potential problems with driving—health (medical conditions and medication use); driving abilities (vision, cognition, and psychomotor); and experiences/attitudes/behavior.

Testing of the workbook indicated that it correlated with an on-road driving test, as well as with several clinical tests of functional abilities, most of which are part of the test battery from the *Model Driver Screening and Evaluation Program*. Older drivers in the validation study considered the workbook useful and reported their intent to make changes in their driving or seek further evaluation as a result of completing it.

In follow-up work to the *Driving Decisions Workbook*, a Web-based self-screening instrument was developed by UMTRI, based on "health concerns" that affect driving—that is, the symptoms that people experience due to medical conditions, medications used to treat them, and the general aging process—rather than the medical conditions or medications themselves. The *SAFER Driving: Enhanced Driving Decisions Workbook* is intended to simplify the self-awareness process, based on the premise that while there are a myriad of medical conditions and medications, they produce a relatively small number of health concerns that vary in severity and in turn affect driving. Testing of the instrument indicated that it correlated with a clinical evaluation and on-road assessment administered through an established driving assessment program at the University of Michigan managed by an occupational therapist and that study participants reported planning to make changes in their driving or seek further evaluation following program completion.

Another home-based awareness tool is the *AAA Roadwise Review*, a CD-ROM based tool adapted from the *Driving Health Inventory* (Staplin & Dinh-Zahr, 2006). Certain procedures were modified but the battery of tests was not changed—thus, the personal awareness tool addresses the same critical safe driving abilities as the clinical tool. Its design was strongly influenced by input from focus groups of older adults. The tool allows older adults to assess their safe driving abilities and also helps them decide how to use the outcomes to continue driving safely by providing links to feedback tailored to individual performance on the measures. An evaluation by Bédard, Riendeau, and Weaver (2007) found that the number of serious problems identified by *Roadwise Review* correlated significantly with problems found during an on-road evaluation.

While early self-awareness results are encouraging, there is clearly a need for further research to evaluate the effects of self-screening on driver behavior. In particular, objective data are needed about the actual changes in behavior made by drivers as a result of self-awareness education (e.g., seeking out further assessment and evaluation, participating in driver education/training activities, and modifying and/ or reducing actual driving), and ultimately whether self-awareness education can help lead to reductions in crash involvement.

FOR THE FAMILY AND FRIENDS

The family and friends of older drivers are often the first ones to notice driving problems. Family and friends are also in a good position to be a trusted source of information and to provide assistance in helping the older driver retain safe mobility. A study of U.S. drivers over age 50 found that more than 63 percent of older adults living with their spouse identified a friend or family member as the person they would most likely listen to about driving concerns (Coughlin et al., 2004; D'Ambrosio et al., 2007). About one-half of older respondents living alone felt this way. A series of focus groups with older drivers and the adult children of older drivers in Michigan found that both of these groups agreed that the families should be the one to talk to the older driver about driving reduction and cessation (Eby, Molnar, Kostyniuk, & Shope, 1999). However, these conversations rarely took place and were generally ineffective when they did. Family members reported that they did not know where to find information on aging and driving.

Fortunately, in the last decade, a number of very good resources have been developed specifically to help families keep their older adults safely mobile (AAA Foundation for Traffic Safety, 2006; AOTA, 2002b; Hartford Financial Services Group, 2007; Land Transport NZ, 2006; New York State Office for the Aging, 2000; Pennsylvania Department of Transportation, 2006; Spreitzer-Berent, 1999; Westchester County Department of Senior Programs and Services, n.d.). Each of the guides provides a wealth of information. Collectively, the guides provide the following general recommendations for families and friends who are concerned about an older driver.

- **Recognize**: Keep in mind that mobility is important for the driver and the driver has unique needs and resources. Talking about driving problems is uncomfortable for drivers, family, and friends alike.
- **Observe**: Look for the types of clues that indicate potential problems, such as stopping in traffic for no apparent reason or other drivers honking their horns. Both Land Transport NZ (2006) and the U.S. Department of

Transportation (2006) provide lists of "warning signs." If you are unsure about warning signs, consult a driving rehabilitation specialist.

- **Communicate**: Be open and honest with the driver about your concerns. Many older drivers would rather be approached by a single family member, usually the spouse, rather than the whole family. The conversation should be direct. Indirect communication, such as hiding the keys, is not effective (Eby et al., 1999). Remember communication involves good listening as well as talking.
- **Encourage**: Formal assessment with a medical professional or an occupational therapist (driving rehabilitation specialist) should be encouraged. Offer to help find the right person and help make the appointment.
- **Assist**: If assessment shows that older adults need to limit or stop driving, assist them in transitioning to nondriving mobility options. This can be as simple as helping them to get around or as complex as developing a transportation plan that includes a list of viable mobility options in the community.

FOR THE HEALTH PROFESSIONAL

Health professionals, such as physicians, nurse practitioners, physician's assistants, and occupational therapists, have an important role in maintaining drivers' safe mobility. Family members of older drivers, state licensing agencies, and the older drivers themselves often seek the advice of a medical professional about aging and driving. A study in the United States found that 74 percent of health-care providers and 83 percent of rehabilitation specialists identified older drivers as one of the three most important areas in which they needed more information (Sterns, Sterns, Aizenberg, & Anapole, 2001). Fortunately, since that study, much more information has been developed on assessing and advising older adults on fitness to drive.

One excellent source of physician education is the previously mentioned *Physician's Guide to Assessing and Counseling Older Drivers* (Wang et al., 2003). Based on review of the literature and expert opinion, the guide is intended to help physicians and other health professionals assess the ability of their older patients to drive safely. The guide provides information about specific red flags for medically impaired driving; a test battery called the *Assessment of Driving-Related Skills* (ADReS) to assess the key areas of function; how to interpret performance on the test battery through scoring cutoffs, as well as examples of interventions to help physicians manage and treat functional deficits identified; driving rehabilitation specialists and how they can be of help; how to counsel drivers who should no longer be driving; physicians' legal and ethical responsibilities; state-by-state licensing requirements, licensing renewal procedures, reporting procedures, and contact information for each state's driver licensing agency and medical review board; and a reference list of medical conditions and medicines that may impair driving skills and consensus recommendations for each regarding driving restrictions. Other outstanding guidebooks for health professionals have been published in Australia (Austroads, 2006) and Canada (Canadian Medical Association, 2006).

REHABILITATION

Rehabilitation refers to the restoration of proper functioning. Some declines experienced by older adults, particularly those related to physical abilities, may be reversible through rehabilitation. While the debilitating effects of medical conditions on driving can also be helped through medical intervention (medicine, surgery, etc.), here we review the research on exercise and cognitive rehabilitation programs.

FITNESS

Fitness training programs involve helping older people drive more safely by improving range of motion, strength, and stamina. These programs show great promise in helping older drivers extend their driving lifetime. Recent work investigated the effects of an exercise program on on-road driving performance (Marottoli, Allore et al., 2007). The program involved an OT visiting older drivers (age 70 and older) weekly for 3 months and guiding them through a graduated exercise program targeting stamina, flexibility, coordination, and speed of movement. Results showed that participants found the program acceptable and maintained their driving performance (as measured through an on-road driving test), while a control group declined in performance.

Another study investigated the effects of an 8-week range-of-motion exercise program on driving performance (Ostrow, Shaffron, & McPherson, 1992). Subjects in the program group (age 60–85) performed the exercises at home and kept a log of their exercise activity. All subjects were tested before, during, and after the program on flexibility and on-road driving performance. The study found that when compared to a control group, participants in the exercise program increased shoulder and trunk flexibility, while flexibility decreased in the control group. The driving performance analysis showed that drivers in the program were significantly better than controls on "observing" (checking mirror, turning to check blind spots) and "vehicle handling" (parallel parking). There were no differences in performance on the other seven driving measures investigated.

Given these promising results and the fact that exercise programs have many other benefits other than improving driving performance, much more research should be conducted in this area. In addition, these results suggest that older drivers who are concerned about declining abilities should consider starting an exercise program that is appropriate for their level of fitness and functioning.

COGNITIVE

One often hears the phrase "use it or lose it" in reference to aging and cognition, and there is evidence to support this assertion (see e.g., Hultsch, Hertzog, Small, & Dixon, 1999). Recent efforts have investigated whether cognitive functioning can be restored through training. There is strong evidence that proper and intensive training can improve cognitive functioning (see e.g., Ball et al., 1988, 2002; Delahunt et al., 2008). The effect of cognitive training on driving performance has also been studied. Kua et al. (2007) concluded that there is only limited evidence that cognitive retraining can improve driving. For example, Roeneker, Cissell, Ball, Wadley, and Edwards (2003) studied the effects of speed-of-processing training and simulator training on driving performance on an open-road test. The experimental group had decreased perceptual/attention functioning while the control group did not. The program used a driving simulator to train older drivers on simple and choice reaction time. Simple reaction time was trained by having the subject brake as quickly as possible in response to simulated brake lights. Choice reaction time was measured by having the subject react to simulated traffic signs. Based on what the sign contained, the subject braked, turned the steering wheel, or did nothing. Results indicated that the experimental group showed improved reaction time after training. As compared to a group of subjects who did not receive training, simulator-trained drivers improved driving performance (measured by on-road evaluation) on only two of the driving measures (turning and signal use) investigated. These improvements, however, had dissipated in an 18-month follow-up. Although the selection of subjects in this study makes it difficult to interpret the results, it seems that more work in this area is warranted. Of particular interest are the products of several companies (e.g., CogniFit, Posit Science, and Nintendo) that show that cognitive functioning can be improved through cognitive training. What is not known is whether these programs can improve cognitive abilities to the level at which they can positively impact the performance of critical driving skills or impact older driver crash risk. More research on cognitive retraining is needed.

CONCLUSIONS

The need for continuing education is emphasized several times throughout this book. Indeed, we propose that mobility education should be a lifelong process. When a child is born, parents should learn how to properly use child safety seats to transport the child, as studies show that the majority of parents misuse them (Eby & Kostyniuk, 1999). As the child outgrows these seats, parents need education to properly use booster seats (Bingham et al., 2006). Once children graduate from booster seats, they need to learn the importance of using their safety belt while traveling in a car, particularly in the back seat (Eby, Kostyniuk, & Vivoda, 2001). Also during this time, parents are teaching their children driving habits and attitudes as the children watch their parents drive and react to traffic situations and other drivers (see e.g., Bianchi & Summala, 2004). When people reach their early teens, they get the most intensive and formal driving education they probably will receive in their lifetime to obtain a driver's license. Interestingly, no studies have found that this education reduces crash risk (see e.g., Mayhew & Simpson, 2002). Once people are licensed they begin to learn about the advantages of independent mobility and the responsibilities that come with it, some learning the hard way through crashes and citations. From young adulthood to old age, drivers improve their critical driving skills, learn various traffic safety messages such as the dangers of impaired driving, learn about the likelihoods of being cited for various traffic violations, and may become dependent on the personal automobile for meeting transportation needs. Sometimes in older adulthood, medical conditions or medications may start to impair a person's ability to drive safely. It is at this point that many people seek, or are forced to receive, the second most intensive mobility education in their life—the type of education described in various places in this book: learning to use community mobility options; learning to use adaptive equipment or advanced technology; learning to drive during safer times and in safer places; learning about new traffic signs and roadway designs; and, most importantly, learning that one day they will need to stop driving. For many of these drivers, it will be the first time they have thought about these issues and how they might meet their mobility needs without driving themselves. While there are many educational needs not being met in our society, society must do a better job of educating drivers at an earlier age about aging and driving.

10 Vehicles and Advanced Technology

It's a never ending battle of making your cars better and also trying to be better yourself.

Dale Earnhardt

INTRODUCTION

Vehicle design, features, and technology are constantly changing. One obvious way to maintain safe mobility in an aging society is to design vehicles and technology that accommodate the age-related changes in functional abilities that tend to be experienced by older adults. Such design changes could help older adults perform critical driving skills that have become difficult for them, while also making the driving task easier for all drivers. In this chapter, we first review the issues surrounding the design of vehicles for older adults. We then discuss how existing vehicles can be modified to help older adults continue to drive safely. We conclude the chapter by discussing new or near-term technologies that have promise for enhancing older driver safety and mobility.

AUTOMOBILE DESIGN

To sell cars, automotive manufacturers design vehicles that are safe, comfortable, and meet the needs of the automobile purchasing public. In 1908, Henry Ford met the needs of the U.S. population by manufacturing the affordable Model T. With a price tag as low as $280, the Model T could be purchased by the average American, allowing most people access to a motor vehicle for the first time (Bellis, 2008b). Vehicle and engine sizes grew to meet the demands of a mostly young public who wanted fast cars, leading to the introduction of "muscle cars" in the 1960s (Wright, 1996). Vehicle accessories have also been offered to entice vehicle buyers. Locking gas caps, for example, first introduced during the depression to prevent fuel siphoning, disappeared after the depression only to reappear during the fuel shortages in the 1970s (Wright, 1996). Vehicle designs are slowly being altered or adapted by automobile manufacturers to make driving easier, more comfortable, and safer for older drivers (Coughlin, 2005; Pike, 2004).

Without a doubt, the automotive industry will respond to the coming wave of aging baby boomers by designing vehicles that take into account the functional declines associated with aging. Current vehicle designs, however, are not meeting these needs. A nationwide study in Great Britain investigated the problems older adults (age 60 and older) had with using cars (Herriotts, 2005). Responses were compared to a younger group who also completed the same questionnaire. The author found that the incidence of self-reported health problems (arthritis, back problems, hearing, heart/circulation) was significantly higher for the older age group. Older adults also reported significantly more difficulty than younger drivers in turning around to look out the rear window (56 percent vs. 33 percent) and getting in and out of the car (32 percent vs. 26 percent). In addition, one-fifth of older drivers had difficulty using safety belts and also reported difficulty taking things in and out of the trunk. Other work has found similar age-related problems with vehicle design (e.g., Murray-Leslie, 1991; Petzäll, 1995; Shaheen & Neimeier, 2001; Zhao, Popovic, & Ferreira, 2007; Zhao, Popovic, Ferreira, & Lu, 2006).

Indeed, the lack of congruence between age-related functional declines and vehicle design has, in part, led to the development of the CarFit program (http://www.asaging.org/carfit). Sponsored by the American Society on Aging, AAA, AARP, and the American Occupational Therapy Association, the nationwide program is designed to help older drivers find out how well they currently fit their personal vehicle, highlight actions they can take to improve their fit, and promote conversations about driver safety and community mobility. CarFit is an educational, hands-on program in which trained instructors discuss with drivers how to make adjustments to their vehicle, such as seat and mirror adjustments, to make driving their vehicle easier and safer. When necessary, the instructors also discuss options for modifying or adapting the vehicle to improve driving. An evaluation of the program in 12 cities found that the majority of drivers needed to make at least one adjustment to their vehicle, reported using safety devices (e.g., safety belts) more often after the program, and were more likely to have conversations with their friends and family about their driving (Finn, 2006).

One factor that may be preventing automotive manufacturers from developing more age-friendly vehicles is the lack of research that has specifically addressed optimizing the design of vehicles to account for age-related functional declines. This sparse literature is reviewed by Shaheen and Neimeier (2001). Based on this literature, the authors proposed vehicle design guidelines for accommodating older drivers. Table 10.1 shows their recommendations by vehicle feature.

Although these recommendations are a good start, they still are not specific enough to help a vehicle-design engineer make decisions. As aptly stated by Zhao et al. (2007), "vehicle innovations for...elderly drivers cannot be based only on the designer's intuitions but must be grounded in users' actual needs and activities. Vehicle designers need to understand what the vehicle means to the new aging population" (p. 6). Much more research is needed in this area.

TABLE 10.1
Recommendations for Changes in Basic Design Features to Help Accommodate Age-Related Declines in Functional Abilities

Vehicle Design Feature	Recommendations
Seats	• Seats should be hard and flat • Seats should not have much friction • Seat position should be easy to adjust • Seat should easily slide back and swivel outward
Doorways	• Width should be 800–900 mm • Sill height should be 50–90 mm • Height should be 1330–1380 mm from ground • Angle 70–90 mm
Safety belts	• Have a specialist adjust the anchorage point and slack to enhance comfort
Handles, knobs, and steering wheels	• Enlarge vehicle knobs, handles, and steering wheels • Knob shapes should be easy to hold, so that it fits in the hand • Texture should be easy to grip and hold • Stop and start should be located apart from one another • Facilitate single-handed task if possible
Antiglare	• Ensure adequate light level on text and controls • Select appropriate color, size, and chromatic intensity for type symbols • Isolate priority information from background clutter and glare • Use appropriate type size and weight and letter, word, and line spacing • Maximize contrast between type and background • Use nonreflective surfaces • Use contrasting colors • Use blue-violet green combinations with care
Driving controls	• If adaptations are made, make sure that secondary driving functions (horn, signals, headlights) are still possible • Use simple extension bars from gear lever if possible • Use lever release mechanisms for hand braking
Auditory information	• Provide a control for sound to ensure that the loudness can accommodate all users • When communication impact is critical, relate the sound volume to the cue urgency and provide both visual and auditory cueing • Minimize ambient noise • Avoid irrelevant information • Use lower frequencies for alarms and urgent messages

Adapted from Shaheen & Neimeier (2001).

CURRENTLY AVAILABLE FEATURES

What should an older driver look for in today's vehicle market? There are vehicle features available today that may make driving easier and safer for older adults. As discussed by Molnar et al. (2007),

- Bright colored cars, especially white ones, may be easier for other drivers to see
- Front- and side-impact airbags provide increased protection
- Firm adjustable seats help with a proper fit and can provide increased visibility and comfort
- Adjustable safety belts are more comfortable
- A steering wheel that adjusts at both the front and at the back allows the driver to find a safe, comfortable position
- Adjustable brake and gas pedals provide for a safer, more comfortable seating position further away from the airbag
- Large windows and small pillar posts provide better visibility by minimizing blind spots
- Controls that are easy to read in any light, reach, and operate help prevent distractions while driving
- Large, glare-proof or self-darkening (light-sensitive) mirrors reduce glare while maintaining visibility
- Wide-angle rearview mirrors and convex side mirrors can be beneficial for those with stiffness in the neck or problems with peripheral vision, but distort distances and require practice to use safely
- Remote keyless entry provides quick and safe access to the vehicle
- Power windows and locks helps prevent distracted driving
- Nontinted windshields make visibility easier

VEHICLE ADAPTATIONS

Vehicle adaptations provide an opportunity for older drivers to compensate for some age-related functional declines that can lead to unsafe driving, such as reduced strength, flexibility, range of motion, and vision-related deficits (Mollenhauer, Dingus, & Hulse, 1995; Mitchell, 1997). Vehicle adaptations help drivers with disabilities and/or aging-related concerns to do things such as get in and out of the car, fasten and unfasten their safety belt, and exert control in operating the vehicle (Shaheen & Neimeier, 2001). Here we review adaptive driving aids that can be added to cars after they have reached the market.

According to NHTSA (2007) there are six steps for older drivers to adapt their vehicle. The first is that they should integrate cost-saving opportunities and licensing requirements. Adapting a vehicle can be costly and the older adult may need a special driver's license, depending upon the adaptations. Monaco and Pellerito (2006) provide examples of strategies to offset the cost of adaptive equipment. Second, they should evaluate their needs. Finding a certified driver rehabilitation specialist (DRS) or OT is recommended. A DRS can be very helpful in making recommendations for adaptive equipment as part of his or her assessment of the impact of functional impairments such as decreased shoulder range of motion due to arthritis, fracture, stroke, or Parkinson's disease. Older drivers not working with an OT may be able to get advice from a rehabilitation agency or hospital about whom to contact to assist them in identifying and obtaining appropriate equipment, installing it, and providing the necessary training. Third, they should select the right vehicle if they are purchasing or leasing one. Not all vehicles can be modified for all adaptations. Fourth, they should find a qualified mobility dealer to modify their vehicle. One resource is to contact the National Mobility Equipment Dealer's Association. Fifth, they should get training on using the new equipment. It is important not only to hear about how to use the technology, but also to practice using the technology on the road with a trainer (Molnar, Eby, & Miller, 2003). Finally, they should maintain their vehicle. It is important to remember that properly modifying a vehicle requires an expert who knows what the specific needs are and how they can be met through adaptive equipment. Improper equipment can increase the chances of getting in a crash.

There are many types of adaptive equipment that can be added to a vehicle. Although adaptive equipment is readily available to the older adult who can afford it, research has shown that many individuals who could benefit from vehicle adaptation, and professionals who work with these individuals, are not aware of the options (Silverstein, Gottlieb, & Van Ranst, 2005). The most common types of adaptive equipment are presented in Table 10.2 by category (Bouman & Pellerito, 2006; Koppa, 2004; Mitchell, 1997). The seating category consists of adaptations to help position the driver safely and comfortably as well as prevent fatigue. The steering category includes a number of devices that can be attached to an existing steering wheel to help the driver steer. The throttle/brake devices either facilitate reaching or pressing peddles, or are adaptations to allow hand control. The items in the auxiliary control category are either modifications to allow a person with limited hand function to use the various secondary features of the vehicle or adapted mirrors to improve visibility for the back and sides of the vehicle.

ADVANCED TECHNOLOGY

In addition to vehicle adaptations, advanced technology systems for vehicles have the potential to increase the safety and mobility of older drivers (Caird, 2004; Perel, 1998). Intelligent Transportation Systems (ITSs) combine advances in wireless communication technologies, automotive electronics, computing, and global positioning systems to help drivers navigate (e.g., global-positioning-based route guidance), see (e.g., infrared vision enhancement systems), and avoid crashes (e.g., collision avoidance systems). These systems vary widely in how they deliver information to the driver and the degree to which they take over control of the vehicle. Other technologies include systems that assist drivers in the event of a crash by signaling for help and showing the vehicle's location.

Successful ITS applications, particularly for older drivers, need to be affordable, be relatively easy to use, and work to enhance safe driving. One way to promote affordability is to develop systems that are flexible enough to benefit drivers of all ages, yet are still able to help older drivers compensate for diminished abilities. The

TABLE 10.2
Commonly Available Adaptive Devices to Facilitate Driving

Category	Vehicle Adaptation
Seating	• Cushions for comfort and boosting the driver for better vision out of the vehicle
	• Custom armrests
	• Safety belt extension
	• Driver side airbag deactivation if the driver needs to sit close to the steering wheel
	• Upper body support
Steering	• Steering knob
	• Spin pin
	• V-grip
	• Palm grip
	• Tripin
	• Steering splint
	• Amputee ring
Throttle/brake	• Left foot throttle
	• Gas pedal block
	• Pedal extensions
	• Hand controls
	• Cruise control
Auxiliary controls	• Adapted wiper, horn, turn signal, cruise control, headlight dimmer controls
	• Push-button ignition
	• Convex and multifaceted mirrors

Adapted from Bouman & Pellerito (2006).

general idea that what works for older drivers will also benefit other drivers is the basis of many successful approaches discussed in this book. However, in the area of ITS, the impacts on driver safety and community mobility, especially for older drivers, are still not well understood.

Until recently, few ITS technologies were developed to take into account the unique requirements of older drivers. In an excellent review of older drivers and ITS, Caird (2004) points out that while older drivers are the group most likely to benefit from ITS, they are also the group most likely to suffer from the potential negative effects of ITS. Poorly designed ITS applications could increase distractions for older users, leading to a higher risk of crash. However, systems designed for optimal use by older drivers would also be beneficial to drivers of all ages. To achieve widespread use of ITS by older drivers, future ITS applications will need to be carefully designed to ensure that safety is enhanced rather than reduced (Henderson & Suen, 1999; Stamatiadis, 2001).

It is clear that older drivers use ITS applications differently than younger drivers (Caird, 2004; Dingus et al., 1997; Eby & Kostyniuk, 1998; Kostyniuk, Streff, & Eby, 1997; Stamatiadis, 1998; Wochinger & Boehm-Davis, 1997). For example, in an evaluation of navigation assistance applications, Kostyniuk et al., (1997) found that older drivers used the system more frequently than young people, entered a greater number of destinations into the system, and utilized the technology with a "copilot." Understanding these patterns of use for the various ITS applications being developed is crucial for optimizing the benefits of ITS for all users (Vrkljan & Polgar, 2007). Such research is lagging. Studies have also found that older drivers take much longer to learn how to use ITS technology (Caird, 2004; Kostyniuk et al., 1997). Whether this is a cohort effect of people who did not grow up using computer technologies or an effect of aging per se is not known. Nevertheless, it is clear that acceptance of ITS applications by older drivers will be largely dependent on the quality of training received.

The most promising ITSs for older drivers include route guidance, night vision enhancement systems, collision warning systems, and automatic crash notification (Caird, 2004). There have been a number of recent projects to develop design guidelines for various ITS applications that take into account the needs and preferences of system users (e.g., Caird et al., 1998; Eby & Molnar, 1999; FHWA, 2002; Kiefer et al., 2003). However, most of the guidelines published from these efforts have focused on the general population and do not specifically address the unique needs of drivers with disabilities and/or aging-related concerns.

ROUTE GUIDANCE

Route guidance systems combine global position system (GPS) vehicle location information with electronic mapping software to provide drivers with real-time instructions to a location as they drive. Route guidance systems have been well researched and several brands, such as Garmin® and TomTom®, are available as an aftermarket addition to vehicles. More advanced systems can also integrate traffic congestion information and help drivers avoid traffic jams.

Route guidance systems can be particularly helpful for older drivers by helping them travel to unfamiliar places; reducing the cognitive workload of reading maps, instructions, and street signs while driving; and increasing feelings of safety while driving. Several studies have examined older drivers' use or potential use of route guidance systems (Dingus et al., 1997; Eby & Kostyniuk, 1998; Eby & Molnar, 1998; Kostyniuk, Eby, Christoff, & Hopp, 1997a, 1997b; Oxley, Barham, & Ayala, 1995; Vrkljan & Polgar, 2007). Collectively these studies show that older drivers

- Use the route guidance systems frequently
- Report only minimal distraction, but more than reported by younger drivers
- Travel to places they would not have without the system
- Report increased feelings of safety, confidence, attentiveness, and relaxation when using the system
- Take much longer to learn how to use the system
- Have more difficulty than younger driver reading the displays

- More frequently use the system with a conavigator than reported by younger drivers
- Would not buy a system targeted to "old" people

Given the fairly low cost of commercially available systems, the positive regard drivers have for them, and the fact that they seem to be safe, route guidance systems are a very promising advanced technology for helping to maintain safe mobility in an aging society.

NIGHT VISION ENHANCEMENT

Night vision enhancement (NVE) systems are designed to provide drivers with roadway information that is either difficult or impossible for the driver to obtain through direct vision. Using technology developed by the military, vision enhancement systems use infrared cameras to detect pedestrians, animals, and the roadway scene. The display is usually placed above the dashboard so that the driver can map the information from the system onto the visual world he or she sees in front of the vehicle. Further description of these systems can be found in Rumar (2002). These systems can already be found in some luxury vehicles as an option.

Studies of NVE systems have utilized both simulators and on-the-road studies (e.g., Druid, 2002; Raytheon Commercial Infrared and ElCAN-Texas Optical Technology, 2000). Collectively, drivers report that they can intuitively interpret the displays; however, this ability seems to be reduced when the display is not positioned above the steering wheel. Research also shows that NVE systems can increase target detection by all drivers and that the system raises driver workload by only a small amount (Sullivan, Bärgman, Adachi, & Schoettle, 2004). Other studies have found that older drivers do not use NVE system displays as frequently as younger drivers, presumably because of decreased divided attention capacity (Gish, Shoulson, & Perel, 2002; Van Wolffelaar & Rothengatter, 1990). Despite this, older drivers report being very satisfied with NVE systems (Ståhl, Oxley, Berntman, & Lind, 1994).

Whether NVE systems improve safety is not yet known. Drivers' comments suggest that NVE systems do not increase distraction, reduce the debilitating effects of headlight glare, and can be easily incorporated into the normal driving routine. On the other hand, the display can be difficult to read for drivers with bifocals (Druid, 2002; Raytheon Commercial Infrared and ElCAN-Texas Optical Technology, 2000). NVE systems may be particularly useful for improving the mobility of older drivers who tend to place nighttime driving restrictions on themselves in response to declining visual abilities at night. However, because of other declining abilities, such as the ability to divide attention, careful research must be conducted to ensure that NVE systems do not increase distraction or workload to unsafe levels with this age group. This research has yet to be done. We conclude that NVE systems are a promising new technology for enhancing older adult safety and mobility.

CRASH WARNING SYSTEMS

A current focus of ITS research is the development of systems that warn drivers of potentially dangerous situations so that they can take appropriate evasive actions or, if appropriate, not perform a maneuver. These technologies, known as *crash warning systems*, use radars and cameras positioned around the vehicle to determine the locations of other traffic and lane pavement markings. When a potentially hazardous situation arises, the system warns the driver. More advanced systems not only warn the driver, but also take over partial control of the vehicle's operation to avoid the hazard. For example, the 2008 models of the Volvo S80, V70, and XC70 have a crash warning system with "auto brake"; that is, the system will detect an impending forward collision and apply the brakes for the driver if he or she fails to do so. There are three main crash warning systems available for automobiles: forward collision warning, intelligent cruise control, and lane departure warning systems.

Forward Collision Warning

Forward collision warning systems use forward radar sensors to determine the distance between cars. When this distance begins to get too small, the system will warn the driver using some signal (usually a combination of a light and sound) and, in some systems, take over partial control of the vehicle. Studies have investigated the safety benefits of forward collision warning systems by placing drivers in simulated conditions where vehicles appeared suddenly in front of them (Cotté, Meyer, & Coughlin, 2001; Kramer et al., 2007; Maltz & Shinar, 2004). These studies have found that driver acceptance was high when the system did not give too many false alarms (giving alerts when they were not appropriate); older drivers were more forgiving of false alarms; older drivers benefited as much as or more than younger drivers; and older participants drove more slowly than younger drivers and maintained longer headways from the next vehicle. In a test-track study, Dingus et al. (1997) found that older drivers maintained longer headways when compared to young drivers. The headways for older drivers did not change as false alarms increased, indicating that older drivers were more tolerant of false alarms. The largest field operational test to date on forward collision warnings system use was conducted by UMTRI and General Motors Corporation (Ervin et al., 2005). In this study, 96 people were given vehicles with a forward collision warning system to drive as they normally would for 1 month. One-third of these subjects were age 60–70. In general, older drivers were likely to view the system favorably and the system improved safety for all drivers. Forward collision warning systems have the potential for extending an older adult's safe driving period.

Adaptive Cruise Control

Adaptive cruise control (ACC), also known as *intelligent cruise control*, is a system that has a forward-mounted sensor that can detect traffic in front of the vehicle, a headway-control algorithm, and an interface with the throttle that can change the vehicle speed to maintain certain headways (Fancher et al., 1998; Hoedemaeker & Brookhuis, 1998). In effect, the driver selects a headway length and the vehicle will stay that time/distance from the vehicle in front without the driver having to use the brake and throttle. A simulator study in Great Britain with 100 participants (age 18–73) investigated how ACC influenced driving behavior (Stanton & Young, 2005). Participants drove a simulated route with various traffic levels with and without ACC. Safety-related variables such as driving speed did not differ when using ACC. The study found that driver workload and stress were reduced when using ACC and drivers trusted the system to work properly. The authors concluded that ACC is beneficial in increasing driver comfort and convenience. A test-track study investigated how ACC affected the driving of 18 participants (age 21–34) in Canada (Rudin-Brown & Parker, 2004). Subjects drove an ACC-equipped vehicle behind a "surrogate" lead vehicle, while performing nondriving tasks. The study found that use of ACC reduced driver workload and that drivers trusted the system even after a simulated failure. Drivers also had greater lane position deviation when using ACC. This latter result suggests that less attention was being paid to steering when using the system, but drivers were not crossing over lane boundaries. ACC use has also been studied under natural driving conditions. Fancher et al. (1998) gave 108 participants (one-third age 60–70) an ACC-equipped vehicle to use as their own for up to 5 weeks. Collectively, these participants used ACC for more than 35,000 miles. The study found that all drivers were overwhelmingly pleased with the system and thought it was trustworthy and safe. The authors reported no crashes during the period of ACC use and, based on several analyses, concluded that ACC is safe. This system is recommended for reducing the driving workload of older drivers.

Lane Departure Warning

Lane departure warning (LDW) systems are designed to help drivers avoid drifting off the road by providing a warning to the driver when the vehicle moves over a dashed or solid lane edge boundary without the use of a lane-change signal (LeBlanc et al., 2006). LDW systems use cameras pointed at the roadway on each side of the vehicle and video-analysis software to determine the vehicle's lane position. Alerts are usually directionally linked so that a drift to the right is accompanied by a light, sound, or seat vibration in the right portion of the vehicle. A simulator study in Germany investigated whether an LDW system could help prevent drowsy-driving crashes (Rimini-Doering, Altmueller, Ladstaetter, & Rossmeier, 2005). After eating a rich meal, 63 healthy young males drove the simulator for up to 2.5 hours on a route with little stimulation. Collectively, the drivers had several hundred microsleep episodes. The study found that when using the LDW system, there was a significant reduction in the number, time, length, and area of lane departure events. There also seem to be safety benefits when LDW systems are used in natural driving. In an investigation of 78 people (26 of whom were age 60–70) in Michigan using a LDW system in over 83,000 miles of driving, LeBlanc et al. (2006) found that the system induced drivers to stay closer to the center of the lane and use their turn signals more often when changing lanes, and reduced the frequency of lane excursions. People also liked the system. We conclude that a LDW system would have great benefit for older drivers, particularly those who are taking medications that can produce drowsiness.

AUTOMATIC CRASH NOTIFICATION

Automatic crash notification (ACN) or mayday systems employ wireless communication technology that will automatically contact emergency medical services personnel in the event of a crash and transmit vehicle location information (Williams, 2002). Some systems can also transmit details about crash severity, giving emergency personnel a general idea of the type of injuries they will encounter (Champion et al., 2003). Usually ACN systems are linked to airbag deployment, but other triggers are available. While not designed to facilitate mobility, ACN systems can aid in saving lives by dispatching emergency assistance earlier than is normally possible. This technology is available from many vehicle manufacturers such as General Motor's OnStar® system. Champion et al. (2003) presented data showing the safety benefits of notifying emergency personnel within the first 10 minutes of a crash (the golden minutes) and transporting an injured victim to the hospital within the first hour of a crash (the golden hour). Several studies have demonstrated the safety benefits and efficacy of ACN systems (Berryman, 2004; Clark & Cushing, 2002; Kanianthra, Carter, & Preziotti, 2000; Ram, Talmor, & Brasel, 2005). No research has directly considered the safety benefits of ACN systems for older drivers, but these systems would undoubtedly be useful for this age group. One concern, however, is that the crash severity and potential injury severity information sent to emergency personnel may not take into account the increased frailty of older adults.

CONCLUSIONS

This chapter has reviewed the various features and technologies that can be added to a vehicle to enhance the safety and mobility of older adults. The automotive industry already includes features in vehicles that facilitate their use and comfort for older drivers, and there are many new technologies on the horizon. It is, however, important to note that these advances in vehicle features and technologies will not reach the majority of the older adult population for some time. According to the 2001 National Household Travel Survey (Hu & Reuscher, 2004), people are owning vehicles longer than in the past. The average age of a household vehicle increased from 5.5 years in 1977 to 9.0 years in 2001, and two out of five vehicles were 10 or more years old in 2001. If these trends continue, it could take more than a decade for the majority of the market to benefit from new vehicle features. In the meantime, aftermarket vehicle adaptation and technology continue to be available for those who can afford them.

11 Roadway Design

Thanks to the Interstate Highway System, it is now possible to travel across the country from coast to coast without seeing anything.

Charles Kuralt

INTRODUCTION

The majority of roadways in the United States and most other countries are more than 50 years old. The U.S. interstate roadway system, for example, celebrated its 50th anniversary in 2006 (Federal Highway Administration, FHWA, 2008b). When these roadways were being built, life expectancy in the United States was only 68 years (Centers for Disease Control and Prevention, CDC, 2003), an age that today is considered to be barely older adulthood. It is therefore not surprising that roadways were generally designed to accommodate the driving capabilities of yesterday's "85th percentile driver"; that is, one who is relatively young and healthy by today's standard (Oxley, Fildes, Corben, & Langford, 2006). Given the types of problems older drivers have on the road, it seems clear that improved roadways can play a key role in enhancing safe driving among older adults.

Intersections are especially dangerous for older drivers (see e.g., Staplin et al., 1998). As described by Dewer (2007), it is possible to reduce the crash risk of older drivers at intersections through changes in roadway design such as protected left-turn signals and improved roadway channeling, stop signs, and signal timing. Similarly, well-maintained roadway markings (e.g., painted edge-lines, lane control marking) can enhance safety by providing visual cues to drivers to help them know which lane to use and to stay in their lane. Some aspects of freeway driving can also be problematic for older drivers—for example, driving through construction zones—and may be made easier by changes in roadway design. Collectively, improvements in roadway design can serve to make the roadway more forgiving not only to older drivers, but also to the general population of drivers on the road. In addition, design improvements at intersections can benefit older pedestrians who are considerably more likely to be killed by automobiles than are younger pedestrians (NHTSA, 2008).

Older adults themselves recognize that the roadways can be improved and report problems with driving that are related to roadway design. Studies in Illinois (Benekohal et al., 1992) and the Netherlands (Meskin, 2002; cited in European Road Safety Observatory, 2006) surveyed older adults about roadway design elements that caused driving problems for them. Significant proportions of older adults reported the following to be problematic.

- Reading street signs
- Driving through an intersection, especially without traffic signal
- Finding the start of a left-turn lane
- Making left turns at intersections, especially without traffic signals
- Following pavement markings
- Responding to traffic signals
- Driving around a roundabout that has more than one lane

Older adults further stated that certain design elements became more important as they aged.

- Nighttime lighting at intersections
- Pavement markings at intersections
- Number of left-turn lanes at intersections
- Width of roadway lanes
- Raised concrete lane guides for turns at intersections
- The size of traffic signals

Others also recognize that roadways should be enhanced for the older driver. A state-of-the-art review of current knowledge and practice to enhance older driver safety was undertaken by the Monash University Accident Research Centre in Australia (Fildes, 1997) to identify best practices. Among other tasks, this project brought together 22 international experts from government, community, and research to prioritize older driver research topics. Of the 15 topics considered, "highway design for older drivers" ranked 4th overall and received a top priority ranking from 10 of the participants. Indeed, several U.S. agencies have recognized the potential of roadway design changes for improving safety and mobility in an aging society (e.g., Potts et al., 2004; Staplin, Lococo, Byington, & Harkey, 2001; Stutts, 2005).

In this chapter we review the following categories of roadway design related to the older driver: traffic control devices, intersections, older pedestrians, and highway work zones. We conclude the chapter by discussing promising approaches to enhancing older adult safety and mobility through roadway design changes. This chapter is not intended to teach readers how to build roadways. Readers interested in this topic should consult civil engineering texts and resources such as the FHWA Web site. The intent of this chapter is to help readers understand the issues and potential solutions in roadway design that can lead to greater driving ease for older drivers and the population in general.

TRAFFIC CONTROL DEVICES

A critical aspect of driving is to visually gather information about the roadway environment. This information is used for current driving, such as keeping a vehicle in the lane or speed control, as well as for appropriately setting expectations for the upcoming roadway, such as curves or changes in the roadway geometry. In addition, the information is used for helping drivers safely negotiate intersections. Much of this information is conveyed through traffic control devices (TCDs). Although

TCDs are ubiquitous on today's roadways, they did not appear until the early 1900s and were not widely used until recently (FHWA, 2007a). Some of the significant advances occurred, not surprisingly, in Detroit, Michigan, which is also known as the *Motor City*. The first centerline was painted in suburban Detroit in 1911; the first STOP sign was placed in Detroit in 1915; and the first three-color traffic light was installed at a Detroit intersection in 1920 (FHWA, 2007a).

The push to install TCDs came mainly from local automobile clubs, who took it upon themselves to make these installations in order to help their constituents drive. Unfortunately, this led to little uniformity in size, shape, or color of TCDs. States soon realized that the lack of uniformity was a growing problem and began efforts to standardize TCDs. Ultimately these efforts led to the first *Manual on Uniform Traffic Control Devices* (MUTCD) in 1935 (Hawkins, 1992). The MUTCD has been continually updated and now appears on the Internet (FHWA, 2007a). Further, standards have been developed to help engineers and planners decide about the use and placement of TCDs to better accommodate older drivers and pedestrians (Staplin, Lococo, Byington, & Harkey, 2001).

While there are a variety of TCDs, they can generally be classified as signs, pavement markings, and signals. Regardless of the type of TCD, it must attract the driver's attention (conspicuity), it must be easily read or interpreted (legibility), it must be understandable (comprehension), and its information must be able to be acquired quickly (Dewar & Olson, 2007). A TCD will not help a driver or pedestrian if he or she does not notice it. Conspicuity is related to the TCD's size, angle of observation from the driver's line of sight, and its color/brightness (Dewar & Olsen, 2007). A TCD's conspicuity can also be affected by environmental conditions, graffiti/vandalism, vegetation coverage, and wear. Conspicuity can be decreased by age-related declines in visual and cognitive functioning, particularly at nighttime. Once the TCD is noticed, the driver or pedestrian must be able to extract or read the information. The ease of reading the sign is known as *legibility*. Several factors influence the legibility of a TCD including the distance from the person, font or symbol size, color, background luminance, and the amount of time the driver has to extract the device's information. Legibility is also influenced by environmental conditions and the visual and cognitive abilities of the driver or pedestrian (Dewar & Olson, 2007). Once the information from the TCD is received, the driver or pedestrian must be able to understand it. TCD comprehension must be quick and correct for the device to be effective. If comprehension takes too long or the information is misunderstood, the driver may not be able to utilize the information when needed, leading to potential safety problems. Several studies have found, for example, that U.S. signage is often misunderstood (see e.g., Ben-Bassat & Shinar, 2006; Dewer et al., 1994; Ogden, Womak, & Mounce, 1990).

SIGNAGE

All of us are aware of roadway signage, particularly when we are traveling in an unfamiliar area. Signs are used to convey information that is critical for the driving task. Dewar and Olson (2007) classified signs as belonging to ten categories. Most of these categories, however, are really subcategories of four basic types of signs (Eby

& Kantowitz, 2006): those that inform drivers of upcoming driver actions (e.g., stop signs); those that warn drivers of possible dangerous situations (e.g., "curve ahead" signs); those that inform drivers of laws and regulations (e.g., speed limit signs); and those that inform drivers of local sites, services, and conditions.

To enhance sign conspicuity, legibility, and comprehension, symbols are often used instead of words. Symbols can convey more information with fewer characters and are legible at further distances (Dewer et al., 1994; Jacobs, Johnson, & Cole, 1975; Kline, Ghali, Kline, & Brown, 1990). Symbols have the additional benefit of being language independent so that nonnative drivers can utilize the signage. However, much of the driving public does not understand what certain symbols indicate (Al-Madani & Al-Janahi, 2002; Dewar et al., 1994; Shinar, Dewar, Summala, & Zakowska, 2003).

In a cross-cultural comparison of traffic sign symbol comprehension in four countries and five driver groups (including older drivers), Shinar et al. (2003) administered a sign symbol comprehension test using signs printed on cardboard cards. Subjects were shown each sign and asked to report what the signs meant. Responses were scored by the experimenters as correct, partially correct, incorrect, or opposite of the true sign meaning. The authors found that in three of the four counties, older drivers (age 65 and older) had the lowest percent of completely correct sign comprehension. Older drivers in Canada and Poland identified sign symbols perfectly less than one-half of the time. When the authors analyzed signs that were specific to the drivers' own country, comprehension improved, but was still less than 70 percent for the older drivers in two of the countries.

A study in the United States and Canada tested comprehension of all 85 symbols in the MUTCD as a function of age (Dewar et al., 1994). Subjects were shown color slides of symbols and were asked to write down what they meant. Answers were scored as correct, partially correct, or incorrect. Drivers age 60 and older had significantly poorer understanding than younger drivers for 33 of the symbols. Ten of the symbols were understood by less than 40 percent of drivers, regardless of age. The authors also presented drivers with 13 symbols modified to improve comprehension and found little improvement in comprehension, especially for drivers age 70 and older. The authors concluded that educational efforts are needed to improve comprehension.

Word signs are by far the most common signs found on roadways and much research has gone into their development. The early work on sign legibility did not consider the diminished visual capacities of older drivers (Forbes & Holmes, 1936), and many signs on the road today are difficult for older drivers to read (Olson, Sivak, & Egan, 1983; Sivak, Olson, & Pastalan, 1981; Staplin et al., 1989). Work by Staplin, Lococo, and Sim (1990) investigated word-sign legibility between young and older drivers in a simulation experiment. Among other variables, the authors varied letter size and found that older drivers needed letters that were as much as 30 percent larger in order to be able to read them as well as younger drivers.

Sign legibility can be adversely affected by darkness, particularly for older drivers. A study by Olson, Sivak, and Egan (1983) conducted in a laboratory where signs were shown to volunteer observers included several variables including glare, background color, background luminance, and legend size. They found that older subjects (average age 68) performed more poorly than younger subjects (average age 24) for most of the variables tested. Research by Dewer, Kline, Schieber, and Swanson (1994) found that sign legibility distances at night for older drivers were about twice those of younger drivers, and three times the distance when glare was present.

Even with good legibility, drivers of all ages sometimes do not understand what the words on a sign mean. For example, a study of 1,745 Texas drivers tested the comprehension of several word signs (Hawkins, Womack, & Mounce, 1993). The authors developed a videotaped survey that was shown to participants in licensing agencies throughout Texas. A sampling plan was used to ensure the subjects were representative of the state. Drivers completed the survey by verbally selecting a response from a set of four choices. Of the 15 word signs tested in the study, older drivers had poorer comprehension than young drivers for six of them, and better comprehension for one (SPEED ZONE AHEAD). Further, drivers regardless of age had poor understanding of many word signs, including LEFT LANE MUST TURN (20.5 percent not understanding); YIELD (20.6 percent not understanding); SPEED ZONE AHEAD (45.0 percent not understanding); and LIMITED SIGHT DISTANCE (55.1 percent not understanding).

Pavement Markings

Pavement markings include painted lines, painted curbs, raised/reflective markers, rumble strips, and word or symbol messages. Markings are used for assisting in lane keeping, channeling traffic, conveying warnings of speed changes, warning drivers about obstacles, and identifying lane-specific characteristics (e.g., turn lane; Eby & Kantowitz, 2006). Pavement markings communicate information through colors (e.g., blue indicates parking for disabled drivers), patterns (e.g., a dashed line indicates that traffic may cross over the line), width (e.g., double wide lines are used for emphasis), and types (including symbols, words, and other special markings) (FHWA, 2007a).

Pavement markings are particularly important at night when they are illuminated by vehicle headlights and ambient lighting. Nighttime conspicuity can also be enhanced by the use of paint with retroreflective properties. (Retroreflective material is designed to reflect light back in the direction of the light source.) Conspicuity and legibility can be reduced due to a number of factors including poor weather, aging or faded paint, and poor lighting at night.

Research has found that pavement markings at night are less visible to older drivers than to younger drivers even with high-beam headlights (Benekohal et al., 1992; Graham, Harrold, & King, 1996; Zwahlen & Schnell, 1999). For example, a study in Ohio examined pavement marking visibility on a fully marked experimental roadway (an abandoned runway) illuminated only by the driver's headlights (Zwahlen & Schnell, 1999). Healthy older adult and young subjects drove slowly along the runway (8–16 km/h) and reported when they saw the end of a finite-length pavement marking treatment. End-detection distance was used as the measure of visibility, with longer distances corresponding to better visibility. The authors varied headlamp illumination (high and low beam) and level of retroreflectivity (medium, high). The study found significantly higher end-detection distances for younger drivers when compared to older drivers. The effect of retroreflectivity level did not affect end-detection distance, but headlamp level had a slight but significant effect on visibility for older drivers. The authors noted that "the small visibility benefit for the older subjects when using high-beam illumination instead of low-beam illumination may be explained in part by the fact that the critical visual detail of the pavement markings at a distance falls below the available contrast sensitivity" (Zwahlen & Schnell, 1999, p.160), suggesting that headlamp difference in marking visibility was related more to the dimensions of the markings rather than to the brightness.

Fortunately, studies have found that enhanced roadway markings improve detection distance and driving performance for both young and older drivers (Horberry, Anderson, & Regan, 2006; Ohme & Schnell, 2001). A field study in Iowa compared the visibility of enhanced pavement markings on a rural two-lane road test site under low-beam illumination (Ohme & Schnell, 2001). Several enhancements were tested including various types of retroreflective material, different marking widths, and "wet weather" tape (a material designed for enhanced visibility on wet roadways). The subjects' task was to drive along the roadway at slow speed (16 km/h) following a pavement marking treatment and to report when they saw a gap in the marking. Subjects were tested in both wet- and dry-road conditions. The authors found no difference in detection distance as a function of marking width, but a significant difference among types of marking materials. The study found that the standard roadway markings had significantly shorter detection distances than the enhanced markings, especially in wet conditions with wet-weather tape. In this condition, there was a 329 percent increase in detection distance over the standard marking for young drivers and a 217 percent difference for older drivers. However, the apparent difference between the age groups was not statistically significant.

Enhanced pavement marking may also improve driving performance during demanding driving situations. An Australian study using a driving simulator tested the effects of enhanced pavement markings on driving performance in wet, night-time conditions, including the placement of cellophane over the simulated vehicle's windshield to simulate rain-induced haziness (Horberry et al., 2006). On one-half of the simulated drives, subjects were required to do a mental arithmetic task intended to cognitively load the driver. Both young and older subjects were tested and several driving performance measures were recorded including mean speed, standard deviation of speed, standard deviation of lane position, and crossing of center/edge lines. The study found that with the enhanced pavement markings, people drove closer to the target speed and were less variable in maintaining this speed, and drivers positioned the vehicle in the roadway more accurately and crossed over line marking only one-half as many times as compared to the standard markings. There were also no age differences found in the study—all drivers benefited from the enhanced markings.

Even when pavement markings are conspicuous and legible, research has found that they are difficult for many people to understand. The study in Texas described earlier found surprisingly poor comprehension of pavement markings (Hawkins et al., 1993). This study looked at seven markings used in Texas. More than one-half of subjects could not correctly select the meaning from a list of four choices of a "single broken white line"; one-quarter of drivers did not understand the meaning of a "single broken yellow center line" or a "solid white edge line." Nearly all drivers knew the meaning of "no-passing zone" markings. There were very few differences by age. Of the seven markings studied, two were understood significantly less by older drivers, one was understood better, and the rest showed no difference. The authors concluded that public information and education programs should be implemented to improve comprehension.

SIGNALS

Signals are used to convey information to drivers at roadways where vehicles come into conflict, such as crossing paths at an intersection. Because they are placed in the driver's field of view and are lighted, signals are generally conspicuous and legible. The common three-color traffic signal is well understood by drivers. Comprehension of other signals, however, may be poor.

Other signals, such as protected left turns and freeway lane control signals, can be difficult for drivers to understand (Hummer, Montgomery, & Sinha, 1990; Ullman, 1993). A study conducted at the 1988 Indiana State Fair assessed drivers' understanding of various alternatives for left-turn signals including permissive (turn when a sufficient gap in the oncoming traffic appears, denoted by a solid green light); protected (exclusive left-turn rights, oncoming traffic is stopped, denoted by a green left-pointing arrow); and a combination of both (solid green light, followed by the green left arrow; or vice versa). Subjects were shown nine sign and signal displays and asked to choose the correct action from four alternatives. The study found that the protected left-turn signal was the best understood while the protected/permissive was the least. The authors found no age differences. Ullman (1993) studied drivers' comprehension of freeway lane control signals (down arrows and Xs) from the MUTCD. For some of the signals, fewer than one-half of subjects knew the proper driving action required by the signal. No age comparisons were presented in this study. A different study of signal comprehension by age, however, found that older drivers (age 60 and older) had poorer comprehension than younger drivers for left-turn signals and emergency flashing modes of traffic lights (Drakopouos & Lyles, 1997). Signal comprehension should be addressed in educational programs for older drivers.

INTERSECTIONS

The intersections of roadways are dangerous places. Intersections involve the potential for vehicle-to-vehicle and vehicle-to-pedestrian conflict, require drivers and pedestrians to respond appropriately to TCDs, have a high concentration of TCDs requiring travelers to process greater amounts of information than for nonintersection locations, provide rapidly changing information as signals and other travelers' positions change, and can be visually cluttered in urban areas. It is not surprising that crashes frequently occur at intersections. According to the FHWA (2007b), intersection crashes account for more than 45 percent of all crashes and 21 percent of fatalities.

Studies show that when compared to younger drivers, older drivers have more difficulty with intersections. A study in Canada used a vehicle instrumented with a forward video camera and a global positioning system (GPS) to objectively measure the location, speed, acceleration, and several driving maneuvers of young, middle-aged, and older-adult drivers (Porter & Whitton, 2002). Participants drove a 26 km route on a mix of local roadways in Winnipeg. The study found that younger participants drove faster, had greater changes in acceleration, and had more video-derived "infractions," that is, errors that would have counted against the driver had an examiner been scoring the drive. Older drivers, on the other hand, had more infractions involving intersections, including improper signaling, improper approach or movement through the intersection, failing to yield, failing to clear the intersection, leaving the turn signal on after turning, and unintentionally running stop signs. This latter infraction happened with two of the 10 older drivers and not at all with the 10 younger drivers.

Edwards et al. (2002) studied the behaviors of young and older drivers at complex intersections using a driving simulator. Participants drove simulated roadways that contained a number of controlled intersections. At random intersections, critical events occurred that required the driver to respond. These events were a pedestrian appearing during a right turn, a last-second yellow light, and a vehicle violating a red light when the participant had a green light (vehicle incursion). Participants were scored based on the speed with which they responded to the event. Young drivers responded faster in all conditions. In terms of the maneuvers performed, the older drivers were less safe. The study found that older drivers ran more yellow lights and tended to not brake for the vehicle incursion event (getting hit by the vehicle on two occasions).

The problems older drivers have negotiating intersections is reflected in crash data. For drivers age 65 and older, 41 percent of fatal crashes are at intersections and this percentage increases with age (IIHS, 2008). Figure 11.1 shows the percent of passenger vehicle drivers in fatal crashes at intersections by age and number of vehicles involved (single vs. multiple). This figure shows that fatal multiple-vehicle intersection crashes increase with age after about age 55, the percentage of fatal single-vehicle intersection crashes does not vary by age, and the vast majority of fatal intersection crashes involve multiple vehicles.

Numerous studies have found that older drivers are over-represented in intersection crashes (e.g., Abdel-Aty, Chen, & Schott, 1998; Baker et al., 2003; Chandraratna & Stamatiadis, 1993; Cook et al., 2000; Garber & Srinivasan, 1990; Hauer, 1988; Kostyniuk, Eby, & Miller, 2003; McGwin & Brown, 1999; McKelvey & Stamatiadis, 1988; Preusser et al., 1998; Rothe, 1990; Ryan, Legge, & Rosman, 1998). A study in the United States, for example, examined the fatal crash risk of drivers age 65 and older relative to drivers age 40–49 during 1994–1995 (Preusser et al., 1998). The study found that drivers age 65–69 were 2.3 times more at risk for a multiple-vehicle fatal crash at an intersection when compared to the younger driver group. Risk of a fatal intersection crash increased with age. Drivers age 85 and older were 10.6 times more likely to have a multiple-vehicle fatal crash at an intersection than were younger drivers. The authors also found that stop-sign controlled or uncontrolled intersections were especially dangerous for older drivers.

What makes intersections so dangerous for older drivers? Two recent studies examined the differences in errors negotiating intersections between younger and older drivers (Braitman et al., 2007; Mayhew et al., 2006). Collectively these studies found that in intersection crashes, older drivers were more likely to fail to yield the

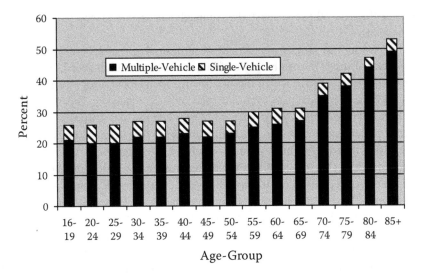

FIGURE 11.1 Percent of passenger vehicle drivers in fatal crashes at intersections by age and number of vehicle involved (single v. multiple). Data are from the Insurance Institute for Highway Safety (2008).

right-of-way, disregard the traffic signal, be responsible for the crash, be at stop-controlled or uncontrolled intersections, and be turning left. This work also found that older drivers tended to be in crashes when roadway conditions were relatively safe, such as during the day and on dry roads. Both studies suggested countermeasures to help reduce the risk of intersection crashes including advanced vehicle technology (such as collision avoidance systems), education and training programs, and intersection modifications, such as the more frequent use of roundabouts.

ROUNDABOUTS

A roundabout (also known as a *rotary* or *ring junction*) is a circular nonsignalized intersection design where all traffic moves in the same direction around the center of the intersection. The history of the roundabout in the United States is turbulent (see Waddell, 2008, for an excellent historical account of the roundabout). Roundabouts were common in the United States in the early 1900s, prior to most cities having traffic signals. The roundabout was ideal for low-speed traffic and moderate traffic volumes. These old roundabouts were governed by a simple rule: Vehicles in the roundabout must yield to vehicles that are entering. As traffic speed and volumes increased in the 1920s and 1930s, the old-fashioned roundabout lost its effectiveness. As traffic volumes increased at roundabouts, vehicles in the roundabouts yielding to incoming traffic (as required) stopped traffic downstream from entering. When traffic volumes reached a critical level, roundabouts would "lock up," dropping the roundabout capacity to zero (Waddell, 2008). Roundabouts quickly fell out of favor and over the next 50 years, most U.S. roundabouts were replaced with signalized intersections.

After World War II, city planners avoided designing roundabouts. At the same time, much of the U.S. road building efforts turned to the high-speed, high-traffic-volume interstate system, which did not involve roundabout design principles. For nearly 50 years, roundabouts were not used on U.S. roadways. During this same time period, however, Europe (particularly Great Britain) and Australia continued to improve roundabout technology, including changing the yielding rule so that vehicles entering the roadway must yield to vehicles in the roundabout. This rule change prevented lockup (Waddell, 2008). Even though research was showing the safety and traffic-capacity benefits in Australia (Daley, 1981; O'Brien & Richardson, 1985), the reputation of the roundabout remained unfavorable in the United States.

Despite this, the modern form of roundabouts was gently introduced back to America in 1990 (Waddell, 2008) and has slowly grown in popularity since then. The negative perception, however, is still present and there is invariably public outcry when new roundabouts are installed (Retting, Kyrychenko, & McCartt, 2007). One reason for increased implementation of roundabouts is that they may help alleviate some of the difficulties older drivers have with negotiating intersections. Several studies have shown that roundabouts reduce the number and severity of crashes (Elvik, 2003; Flannery, 2001; Flannery & Datta, 1996; Oxley, 2006; Persaud, Retting, Garder, & Lord, 2000). Elvik (2003), for example, found that changing signalized intersections to roundabouts can reduce the total number of injury crashes by up to 50 percent and fatal crashes by up to 70 percent. These safety benefits were found for drivers of all ages.

Are roundabouts safer for older drivers? Studies show that older drivers are concerned about negotiating roundabouts (Benekohal et al., 1992; Lord, Schalkwyk, Staplin, & Chrysler, 2005; Meskin, 2002), particularly ones that have multiple lanes. It is likely that much of this concern stems from a lack of familiarity with the roundabout design, rules, signage, and pavement markings. Lord, Schalkwyk, Chrysler, and Staplin (2007) have provided design guidelines that may improve older drivers' comfort, confidence, and safety with negotiating roundabouts. Although we could find no study that specifically evaluated the safety benefits of older versus younger drivers using roundabouts, we conclude that roundabouts improve the safety of older adults in the same way they improve traffic safety for drivers of all ages. Because of the lack of familiarity for U.S. drivers, roundabout public information and education should accompany installation.

Because roundabouts do not have stop signs and traffic lights, some people have expressed concern that they might pose a risk for the safety of pedestrians, particularly for those who are visually impaired or have mobility difficulties (Wall et al., 2005). Although the data are sparse for the United States, research in Australia (Jordan, 1985) and Scandinavia (Ulf & Jorgen, 1999) have shown that roundabouts are, in fact, safe for pedestrians. A case study in the United States, using a variety of converging methods to determine roundabout pedestrian safety, concluded that roundabouts will improve pedestrian safety (Stone, Chae, & Pillalamarri, 2002).

PEDESTRIANS

The safety of older pedestrians is of concern. Figure 11.2 shows the 2006 population-based fatality rate per 100,000 people in the United States by age group (NHTSA, 2008). The fatality rates for the oldest age groups are higher than for other age groups. In 2006, more than 900 pedestrians age 60 and older died in the United States (NHTSA, 2008).

Given what is known about age-related declines and the fact that most older adults prefer to travel in an automobile, older pedestrians are likely to have perceptual, cognitive, or psychomotor declines that make it difficult for them to safely walk along roadways (Langlois et al., 1997). A detailed analysis of roadway design features and older-pedestrian crashes provides some insights into pedestrian road-crossing behavior (Shankar, Sittikariya, & Shyu, 2006). The authors applied multivariate analysis techniques to a pedestrian crash database to control for vehicle volumes. The study found that older pedestrian crashes were more frequent when a center turn lane was present, the spacing between controlled intersections was more than ½ mile, and roadways were poorly lit. While the last finding is obvious, the authors suggested that the first two factors induced older pedestrians to cross roadways at midblock, where there is no traffic control.

Another study of older pedestrian safety at intersections considered vehicle-pedestrian crashes in six U.S. cities during 1995–1999 to determine the effects of marked crosswalks (Koepsell et al., 2002). After controlling for pedestrian flow, vehicle flow, crossing length, and signalization, the study found that intersections with marked crosswalks had more than twice the risk of a vehicle-pedestrian crash. The authors reported that the majority of this increased risk was for marked crosswalks that had

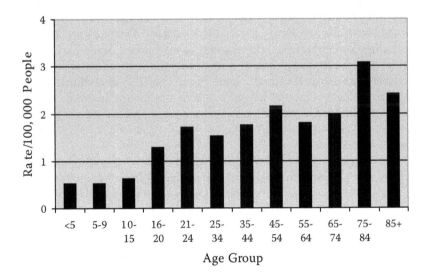

FIGURE 11.2   2006 population-based pedestrian fatality rates per 100,000 people in the United States by age group. Data are from the National Highway Traffic Safety Administration (2008).

no stop sign or traffic signal. This counterintuitive finding suggests that (1) marked crosswalks may give pedestrians a sense of safety or confidence in crossing a road; (2) drivers may not pay attention to crosswalk markings unless there is traffic control; (3) civil engineers should be cautious about installing crosswalk marking when there is no traffic control, especially when older adults may be present; and (4) older pedestrians should be extra cautious when using crosswalks with no traffic control.

ROADWAY WORK ZONES

Although work zones are not a roadway design element, they are common, necessary, and intermittent situations that must be dealt with by drivers. As described by Dewar and Hanscom (2007), work zones can be hazardous "because motorists are confronted with unexpected and often confusing conditions" (p. 403). The most current data show that more than 1,000 people died in work-zone-related motor vehicle crashes in 2006 with another 40,000 being injured (Workzonesafety.org, 2008).

A key feature of work-zone safety is driver expectancy (Pietrucha, 1995). Drivers expect the roadway to be free of obstacles and to be able to travel at the speed of traffic flow. At the same time, a well-controlled work zone will get the driver's attention and meet his or her expectations about how to navigate the work zone (Alexander & Lunenfeld, 1986). The most frequent contributing factor to work-zone crashes is driver inattention and failure to yield the right-of-way (Pigman & Agent, 1990).

Work zones may be especially difficult for older drivers. As discussed in Chapter 2, older people tend to process information more slowly, have more difficulty dividing attention, have vision problems (especially at night), and have slowed reaction times. Other work has shown that older adults respond more slowly to novel stimuli than do younger people (Hoyer & Familant, 1987). Drivers with early stage dementia may be at particular risk because of memory problems. All of these declines can impact older driver safety in work zones. Some work has shown that older drivers may be less safe in work zones than younger drivers. A survey of AARP members across the United States (mean age 72.2) showed that 20 percent of older respondents reported having problems negotiating highway work zones (Knoblauch, Nitzburg, & Seifert, 1997). The specific problems identified by these drivers were congestion, lack of adequate warning, narrow lanes, lane closures/shifts, and lane keeping.

Chiu, Mourant, and Bond (1997) investigated the driving performance of older and younger drivers in a driving simulator. Participants drove through two simulated work zones (all lanes shifted, single lane closed) under three levels of illumination (day, dusk, and night). The authors found that all drivers had more difficulty lane keeping in the dusk and nighttime illumination conditions. All drivers reduced their speed in the nighttime conditions. Older drivers, however, had significantly delayed lane changes in response to the work zones when compared to the younger drivers and this difference was greatest during the nighttime conditions. Given these results, we would expect that older drivers would be over-represented in work-zone crashes. We could not, however, find any studies that have compared the rates of work-zone crashes by driver age.

PROMISING APPROACHES

A recent review of strategies for promoting lifelong community mobility for older adults proposed five components to a promising approach for roadway design (Molnar et al., 2007). The first component was that design guidelines should be responsive to the needs of older drivers. Design standards must take into account the specific driving-related declines that can occur with aging and how these declines impact the ability of older drivers to negotiate the roadway. The second component was that there should be a uniform set of standards that can guide the design of new roads and redesign of existing roads across states and local communities. This is important not only so that drivers find consistency in the designed roadway environment regardless of where they are traveling, but also so that valuable resources are not wasted by having to start from the beginning each time design solutions are needed. The third component was that roadway designers and civil engineers at the federal, state, and local levels should be aware of the standards and understand when and how standards should be implemented, as well as have sufficient resources for actual implementation. To facilitate the training of transportation planners, engineers, and other practitioners, FHWA offers a 1-day training workshop and literature (FHWA, 2003) to assist transportation professionals in making decisions about roadway design improvements. The fourth component was that the strategy should have demonstrated safety benefits and be cost effective. Finally, there must be a sound plan in place to educate older adults about the improvements in roadway design so they will understand the design changes and be likely to appreciate and take advantage of these improvements.

The U.S. FHWA has recognized the benefits of designing roadways to accommodate the older adult population and commissioned *Guidelines and Recommendations to Accommodate Older Drivers and Pedestrians* (Staplin et al., 2001). This document provides numerous guidelines on redesigning roadways to make driving easier for older drivers, based on empirical data about older adult capabilities. This document contains several design features that have potential for enhancing older adult mobility. Each design recommendation is based on what is known about age-related declines in driving and extensive background material on the rationale and supporting evidence for each recommendation. There is also a section intended to help designers and engineers decide when to implement the recommendations. A three-part process is presented that includes problem identification, identification of handbook applications, and implementation decision.

Reproducing the more than 100 specific design elements recommended in the report is beyond the scope of this chapter and of limited usefulness to general readers. However, it is useful to think about the broader strategies that have likely helped to shape some of the specific design elements. The *2004 NCHRP Report 500* (Potts et al., 2004) highlighted promising strategies to improve the roadway/driving environment to better accommodate the special needs of older drivers and characterized each strategy as proven (those strategies that have been used in one or more locations and subjected to properly designed evaluations that show it to be effective), tried (those strategies that have been implemented in a number of locations that may even be accepted as standards or standard approaches, but lack valid evaluations), and experimental (those strategies that have been suggested and found sufficiently promising that at least one agency has considered trying them on a small scale in at least one location). Here is a sampling of proven and tried strategies.

- Provide advance warning signs to inform drivers of existing or potentially hazardous conditions on or adjacent to the road
- Provide advance guide signs and street name signs to give older drivers additional time to make necessary lane changes and route selection decisions, and reduce or avoid excessive or sudden braking behavior
- Increase size and letter height of roadway signs to better accommodate reduced visual acuity of older drivers
- Provide longer clearance intervals at signalized intersections to accommodate slower perception-reaction times of older drivers
- Provide more protected left-turn signal phases at high-volume intersections to avoid difficulties older drivers have with determining acceptable gaps and maneuvering through traffic streams when there is no protective phase and understanding the rules under which permitted left turns are made
- Provide offset left-turn lanes at intersections to reduce potential for crashes between vehicles turning left from a major road and through vehicles on the opposing road because of blocked views
- Improve lighting at intersections, horizontal curves, and railroad grade crossings to help older drivers compensate for reduced visual acuity and provide additional preview distance and more time to prepare for planned actions.
- Improve roadway delineation so older drivers have better visual cues to recognize pavement markings along the roadway and provide raised channelization at intersections to enable them to maintain their lane and safely negotiate through an intersection
- Replace painted channelization with raised channelization to give drivers better indication of the proper use of travel lanes at intersections by providing better contrast and help drivers detect downstream geometric features such as pavement width transitions, channelized turning lanes, and island and median features
- Reduce intersection skew angle to lessen amount of head and neck rotation required of older drivers and provide a wider field of view for the driver to recognize conflicts and select appropriate gaps (by meeting a 90-degree angle or being skewed as little as possible)
- Improve traffic control at work zones to improve driver expectancy by providing adequate notice to drivers describing the condition ahead, the location, and the required response

The report also describes for each strategy, the rationale, target audience, available information on expected effectiveness, keys to success, potential difficulties in implementation, appropriate measures of success, organizational and policy considerations, training and personnel needs, legislative considerations, expected time frames for implementation, and costs.

Alongside the federal efforts to improve roadway design have been local efforts focusing on a limited number of roadway features. Demonstration projects are under way in three states—Arizona, Massachusetts, and Washington—to evaluate the effectiveness of selected recommendations from the report. Results are pending. The FHWA has also sponsored work in Florida to evaluate the effects of report recommendations for intersection design on the driving performance of young and older drivers, using both on-road driving evaluations and a driving simulator. Results are very encouraging but need to be replicated and expanded (Shechtman et al., 2007; Classen et al., 2006, 2007). Another noteworthy project is the AAA Michigan Road Improvement Demonstration Program, which was initiated in 1996 to reduce the frequency and severity of crashes at high-risk urban intersections (Zein & Mairs, 2002). Although it focused on all drivers, it targeted a roadway feature—intersections—that is particularly problematic for older drivers and therefore has the potential to provide increased benefits for older drivers. Partnerships of private and public sector representatives were set up in two urban areas of Michigan (Detroit and Grand Rapids) to plan and implement low-cost safety improvements to traffic signals, pavement markings, and signs (e.g., creating left-turn lanes and left-turn signals, increasing the diameter of traffic signal lenses, and increasing traffic flow by improving light timing). AAA reported preliminary reductions in crashes of 26 percent.

A number of other more general state practices for improving roadways for older drivers are delineated in *NCHRP Synthesis 348* (Stutts, 2005). The report also includes survey results from 24 states relative to modifications they have made to their guidelines or standards for each of 13 engineering treatments to better accommodate the needs and capabilities of older road users. Thus, it is instructive to review the report for more detailed descriptions of promising approaches at the state and local levels.

CONCLUSIONS

Improving roadway design elements to meet the needs and capabilities of older drivers has excellent prospects for improving the safe mobility of everyone. It is, however, impractical to make a wholesale change in roadways—it would simply be too expensive and too disruptive. Instead, we support the guidelines and recommendations in Staplin, Lococo, Byington, and Harkey (2001), and encourage local jurisdictions to consider them when building new roadways, replacing roadway segments, or redesigning intersections that have demonstrated crash problems. We also encourage jurisdictions to engage the older adult public in educational activities. These activities benefit not only older drivers by making them more familiar with design changes, but also the local jurisdictions who will be able to address design misconceptions before they become a problem.

12 Driving Retirement

Any man who can drive safely while kissing a pretty girl is simply not giving the kiss the attention it deserves.

Albert Einstein

INTRODUCTION

Each year, an estimated 600,000 adults age 70 and older stop driving in the United States and become dependent on others to meet their transportation needs (Foley, Heimovitz, Guralnik, & Brock, 2002). Currently, about one in five adults age 65 and older does not drive (Kochera, Straight, & Guterbock, 2005). Having to stop driving can be traumatic and life changing. It is clear from a broad range of studies that reducing and particularly stopping driving are stressful experiences for many older adults, with adverse consequences for their psychological outlook and quality of life (Whelan et al., 2006). Loss of driving can lead to increased social isolation by preventing regular contact with friends and family (Liddle, McKenna, & Broome, 2004; Ragland, Satariano, & McLeod, 2004), and is associated with not only a loss of independence, mobility, and freedom (Adler & Rottunda, 2006; Bauer, Rottunda, & Adler, 2003; Cornoni-Huntley, Brock, Ostfeld, Taylor, & Wallace, 1986; Dobbs & Dobbs, 1997) but also feelings of diminished self-worth, reductions in self-esteem, and loss of identity (Eisenhandler, 1990).

The association between driving cessation and increased depressive symptoms over time has been documented in several recent studies (e.g., Fonda, Wallace, & Herzog, 2001; Marottoli et al., 1997; Ragland et al., 2005). For example, Fonda et al. (2001) examined the effects of driving reduction and cessation using data from a nationally representative panel of adults age 70 and older. Based on interview data with 5,239 older adults (3,543 of them drivers), the authors found that respondents who had stopped driving were more likely to report worsening depressive symptoms, based on the abbreviated Center for Epidemiologic Studies-Depression (CES-D) scale. Having a spouse available to drive them did not mitigate the risk of worsening symptoms. Ragland et al. (2005) found similar results in a longitudinal study on age-related changes in physical activity and function among adults age 55 and older. At baseline, former drivers reported higher levels of depression than active drivers even after controlling for age, gender, education, health, and marital status. In addition, respondents who stopped driving during the 3-year study reported higher levels of depressive symptoms, based on the CES-D, than respondents who continued to drive, even after controlling for changes in health status and cognitive function.

Given these findings, it is not surprising that many older adults are reluctant to stop driving and few plan for the time when they will no longer be able to drive (Bailey, 2004; Kostyniuk & Shope, 2003). However, at some point most older adults will be faced with the necessity of having to reduce or stop driving. This section examines the reasons that older drivers reduce or stop driving, the process they go through, and the efforts that have been undertaken to help older adults transition smoothly from driving themselves to relying on other community mobility options.

REASONS FOR DRIVING CESSATION

The process of driving cessation is complex and decisions to stop or reduce driving are influenced by a variety of factors (Dickerson et al., 2007). Many of the reasons given by older adults for driving cessation relate to health and medical problems, especially vision (Dellinger, Sehgal, Sleet, & Barrett-Connor, 2001; Rabbitt, Carmichael, Jones, & Holland, 1996; Ragland et al., 2004) and to a lesser extent conditions such as Parkinson's disease and stroke-related paralysis (Campbell, Bush, & Hale, 1993). However, the relationship between medical conditions and driving cessation is not straightforward. While Dellinger et al. (2001) found medical conditions to be the most commonly reported reason for driving cessation in a mail survey of 1,950 adults age 55 and older in Southern California, study participants who had stopped driving had fewer medical conditions than did participants who continued to drive. At the same time, former drivers had lower levels of self-reported health than current drivers did, leading the authors to conclude that a broader measure of general health or functional ability may play a large role in decisions to stop driving.

Similarly, in one of the few longitudinal studies of driving cessation, Anstey, Windsor, Luszcz, and Andrews (2006) found that among current older drivers, measures of subjective health and cognitive function were more important than medical conditions for predicting driving cessation. The study involved five yearly waves of data collection (including interviews and clinical assessment of vision, hearing, cognitive function, and grip strength) among 1,466 adults age 70 and older in Southern Australia. The most reliable health-related predictor of driving cessation was found to be self-rated health. In addition, driving cessation was associated with increasing age, low grip strength, and poorer cognitive performance (processing speed, verbal ability, and memory). The authors concluded that subjective feelings of health and well-being may be more important than objective experience of chronic medical conditions and sensory impairment for predicting driving cessation.

Poorer cognitive performance, as measured by processing speed, was also found to play a role in driving cessation in a recently reported longitudinal study by Edwards et al. (2008). The study examined physical, visual, health, and cognitive abilities of 1,656 adults age 65 and older as prospective predictors of self-reported driving cessation over a 5-year period. Cognitive speed of processing, older age, poorer health, and poorer physical functioning were found to be significant risk factors for driving cessation, after taking into account baseline driving. Visual acuity was not a significant risk factor after adjusting for health and physical or cognitive performance.

Several studies have examined the role of social influences, as well as personal and environmental resources in driving cessation (e.g., Johnson, 1995, 1999). The evidence regarding social influences on driving cessation is mixed, with some studies reporting that friends and family do influence older drivers' decisions (Johnson, 1995, 1999) and another reporting no such influence (Persson, 1993). Financial resources may also have an impact on driving decisions; for example, for those older adults living on fixed incomes, the costs of owning, maintaining, and operating a motor vehicle can be prohibitive. However, there has been surprisingly little research addressing the psychosocial influences on this decision, including the role that family, friends, and professionals play in this process (Dickerson et al., 2007).

It is widely accepted that women are more likely than men to stop driving and to do so at an earlier age. Unsworth, Wells, Browning, Thomas, and Kendig (2007) examined a sample of 752 adults age 65 and older from a longitudinal study in Australia conducted from 1994 to 2000 and found that women were three times more likely to give up driving than were men, even when health and disability were taken into account. Other studies have shown that older women voluntarily stop driving at younger ages and in better health than older men, often prematurely (Davey, 2007; Hakamies-Blomqvist & Wahlström, 1998; Siren, Hakamies-Blomqvist, & Lindeman, 2004). In addition, it has been reported that compared to older women, older men make less drastic changes to their driving habits as they age (Eberhard, 1996) and are more reluctant to give up their driving privileges (Kostyniuk, Trombley, & Shope, 1998).

However, several recent studies have failed to find gender to be a significant predictor of driving cessation or have yielded mixed results. Results from the longitudinal study by Edwards et al. (2008), reported earlier, indicated that gender was not a significant predictor of driving cessation after baseline driving was taken into account. The authors concluded that although older women from contemporary cohorts drive less at baseline, they may not be more likely than men to cease driving across time. In another longitudinal study by Anstey et al. (2006), women were significantly more likely than men to stop driving after 2 or 3 years but stopped driving at a similar rate to men 1 year postbaseline. At baseline, nondrivers were more likely to be female. Dellinger et al. (2001) did not find statistically significant differences between men and women in driving cessation overall or with respect to number of miles before stopping, number of medical conditions, or the number of crashes in the previous 5 years before driving cessation.

A possible explanation for these inconclusive findings relative to gender comes from work by Hakamies-Blomqvist and Siren (2003). These authors suggested that while strongly correlated with gender, the decision about driving cessation is related to personal driving history rather than gender per se. The authors surveyed 1,494 Finnish women who had either given up or renewed their license at age 70 and found that the length and level of activity of personal driving history were strongly associated with driving cessation, with former drivers more likely to have an inactive driving career behind them and continuing drivers more likely to have an active personal driving history. In addition, women with an active "male-like" driving history who had decided to stop driving gave reasons for driving cessation that were similar to what is known about older men's reasons for giving up driving.

In interpreting these findings, Siren and Hakamies-Blomqvist (2005) noted that older women's mobility problems result from life course choices rather than simply the aging process. That is, women make choices throughout their life course that affect their driving travel patterns, with the end result for many women being that their driving experiences become quantitatively and qualitatively more limited than men's and they often stop driving earlier than men. However, these differences between men and women are likely to become much less noticeable in future cohorts of older drivers, given that the baby boomers, both men and women, have literally grown up with the automobile (Molnar & Eby, in 2008b) and by 2030, almost all older adults will have been licensed drivers for most of their lives (Rosenbloom, 2004). Over 90 percent of baby-boomer women are licensed, and if baby-boomer women retain their licenses into old age at the same rate as men do now, 84 percent of women age 75 and older will be licensed by 2030 (Spain, 1997).

Despite these trends, gender differences remain an important issue in discussions of safe mobility for several reasons. First, there is considerable evidence that men and women give different reasons for driving cessation (Dellinger et al., 2001; Hakamies-Blomqvist & Wahlström, 1998). Dellinger found that women were more likely to report licensing problems, costs of keeping an automobile, and the availability of someone to drive them as reasons for driving cessation. Hakamies-Blomqvist and Wahlström (1998) also found different reasons among men and women for not renewing their license—men referred most often to health reasons while the most frequent reason for women not to renew their license was that they had already stopped driving. Second, women have different travel patterns and mobility needs compared with men despite discussions often treating men and women as a homogeneous group (Whelan et al., 2006). Third, while older women are driving in increasing numbers, older women in the future will probably continue to be less likely to drive than older men, although the gap will be considerably less than it has been (Burkhardt & McGavock, 1999), and will therefore have a greater need for other community mobility options.

THE PROCESS OF DRIVING CESSATION

Driving cessation is an individual process that is influenced by a host of factors. Some drivers stop driving suddenly because of a medical problem or other precipitating event. In a study by Stutts, Wilkins, Reinfurt, Rodgman, and Van Heusen-Causey (2001), almost three-quarters of participants reported stopping driving suddenly due to health problems, involvement in a crash, or revocation of their license. In a qualitative study using in-depth interviews with women age 65 and older who had stopped driving, Bauer et al. (2003) found that for situations in which there was a precipitating event such as a heart attack or fall, driving cessation was generally sudden. Drivers with dementia may be especially vulnerable to a sudden end to driving. For example, Adler and Kuskowski (2003) surveyed 53 older adult men with dementia as well as a family member or someone close to them. Their findings suggested that the cessation decision was usually unplanned, abrupt, and involved input from physicians and family, although a surprising number continued to drive well after symptom onset.

For many drivers, however, driving cessation unfolds as a gradual process as they become increasingly more vulnerable to difficulties in traffic, limit their driving under certain conditions, and drive progressively less than before (Hakamies-Blomqvist & Wahlström, 1998). Some recent work has focused on identifying broad stages of driving cessation, while recognizing the highly individualized nature of this process. Liddle and her colleagues (e.g., Liddle et al., 2007; Liddle et al., 2004) identified four general stages of driving cessation experienced by older adults including driving in the past, predecision, decision, and postdecision. The driving in the past stage includes the time up to when driving difficulties began and is seen as a highly valued part of one's life. This stage is associated with many important milestones and life roles such as getting a license, use of an automobile for work, and leisure related to travel. The predecision stage is characterized by increased difficulty with driving due to natural changes associated with aging or with the onset of injury or illness, but with no plan to cease driving. The decision stage is characterized by either a voluntary or involuntary decision to stop driving, as a result of either a gradual and careful process of considering options or a short enforced consideration following what can be a traumatic trigger (such as onset of illness or involvement in a crash). In the postdecision stage, retired drivers must make both practical and emotional adjustments to their lifestyle (e.g., finding alternative transportation options and coming to terms with the losses represented by driving cessation, respectively).

SELF-REGULATION BY OLDER DRIVERS

While we still have much to learn about the driving cessation process, we know there is considerable variation in how older drivers respond to driving-related problems, what steps they take to continue driving safely, and how well they adapt if they are forced to stop driving. For example, as part of the driving cessation process, many drivers with functional declines restrict their driving to circumstances under which they feel safest, but others do not appear to practice appropriate driving self-regulation (Baldock, Mathias, McLean, & Berndt, 2006; Charlton et al., 2006; Stalvey & Owsley, 2000). A review of the literature on self-regulation and older driver safety was recently conducted by Molnar & Eby (2008c). In this review, findings from several studies indicated that older drivers self-regulate by reducing their driving exposure (e.g., Benekohal, Michaels, Shim, & Resende, 1994; Charlton et al., 2006; Klavora & Heslegrave, 2002; Marottoli et al., 1993; Raitanen, Tormakangas, Mollenkopf, & Marcellini, 2003; Ruechel and Mann, 2005). There was also evidence that in general, older drivers self-regulate by avoiding specific driving situations such as driving at night, in bad weather, in heavy traffic, and making left turns (e.g., Baldock et al., 2006; Ball, Owsley et al., 1998; Benekohal et al., 1994; Charlton, Oxley, Fildes, & Les, 2001; Charlton et al., 2006; Hakamies-Blomqvist & Wahlström, 1998; Klavora & Heslegrave, 2002; Kostyniuk & Molnar, 2005; Ruechel & Mann, 2005; Stalvey & Owsley, 2000). However, there was considerable variation across these latter studies, making the findings less conclusive relative to the widespread adoption of such self-regulatory practices. Rates of self-reported avoidance of night driving, for example, fluctuated considerably (e.g., 8 percent, Baldock et al., 2006; 25 percent, Charlton et al., 2006; 60 percent, Ruechel & Mann, 2005; and 80 percent, Ball, Owsley et al., 1998).

Mixed results were also found in the review with regard to the association between self-regulation by older drivers and declines in functional abilities. A study by Ball et al. (1998) examined self-reported driving avoidance in a sample of 257 older drivers age 55 and older. The authors found that subjects with clinically determined visual and/or attentional impairments reported avoidance of challenging situations (e.g., driving on high traffic highways, in rush hour, in rain, and alone), although subjects with impaired mental status did not appear to self-regulate their driving. However, findings from a telephone survey of 404 high-risk drivers (defined as those with visual acuity and/or visual processing deficits, a high level of driving exposure, and a history of crash involvement) age 65 and older indicated that most did not acknowledge their visual impairment, more than 75 percent did not self-regulate by avoiding driving situations that placed the greatest demand on visual processing abilities, and the majority rarely performed specific alternative driving strategies (Stalvey & Owsley, 2000).

Baldock et al. (2006) examined self-regulatory practices among 90 drivers age 60 and older in South Australia. Findings indicated that poorer performance on an on-road driving test was not related to overall avoidance of difficult driving situations. However, an association was found for three specific situations—driving in the rain, driving at night, and driving at night in the rain. The authors concluded that older drivers do appear to self-regulate in a manner consistent with driving ability, but only for a small number of specific situations in which they have low confidence and are most able to avoid.

Charlton et al. (2001), in a preliminary investigation of self-regulation among a small sample of older drivers in Australia, compared self-reports from subjects about their driving with results from functional and on-road assessments. They found that overall, self-regulation was associated with poorer levels of functional ability, suggesting that at least some drivers with impaired visual, cognitive, and psychomotor abilities did self-regulate their driving, but that self-regulation was not associated reliably with driving performance as measured on the on-road driving assessment. In more recent work, Charlton et al. (2006) examined the prevalence and type of self-regulation among 656 older drivers age 55 and older in Victoria, Australia, using a telephone survey. Findings indicated that self-reported vision problems were associated with avoidance but decision making was not, and that driving confidence was strongly predictive of avoidance behavior.

Findings relative to the relationship between self-regulation and gender were generally consistent, with women being more likely to report self-regulation than men in the studies reviewed (e.g., Charlton et al., 2006; Hakamies-Blomqvist & Wahlström, 1998; Kostyniuk & Molnar, 2005, 2007). Similarly, it appeared that awareness of and insight into functional impairments was an important precursor to adopting self-regulatory practices (e.g., Ball, Owsley et al., 1998; Freund, Colgrove et al., 2005; Owsley, McGwin, Phillips et al., 2004; Owsley et al., 2003; Stalvey & Owsley, 2003). In addition, self-perceptions of confidence in specific driving situations were closely tied to self-regulation in terms of avoiding those situations (e.g., Baldock et al., 2006; Charlton et al., 2006).

SMOOTHING THE PROCESS OF DRIVING RETIREMENT

Planning for driving cessation needs to occur early, before a health or other crisis occurs that requires immediate intervention without time to think carefully about possible courses of action. Given that many individuals move toward driving cessation over a process that takes years—there is often sufficient time for older adults, their families, and the professionals treating them to discuss and prepare for the transition from driving to nondriving. Yet many of the stakeholders in this process report a lack of direction and support for managing driving cessation (Liddle & McKenna, 2003). Recent work by Molnar, Eby, St. Louis, and Neumeyer (2007) outlined the general components that should be included in efforts to smooth the transition from driving to nondriving, and identified specific programs and initiatives that appear especially promising, particularly in North America. Components of a promising approach included

- Program development/design based on age-related research and broader research on life transitions
- Early intervention and planning to help manage the transition
- Involvement from a broad spectrum of professionals and family
- Recognition that transitioning from driving to other transportation options is a process and an individualized one at that
- Availability of alternative transportation options so older adults have something to which they can transition

According to the authors, effective efforts to help older adults transition from driving should consider what is known about the driving reduction and cessation process, as well as what is known about successful transitioning in other areas of life. As an example, general research on stress and coping suggests that coping with stressful events can be influenced not only by the nature of the stressor, but also by the coping resources on which individuals draw, such as social support, a sense of optimism, and finances. There is also evidence that certain coping strategies may have potential for helping older adults mitigate depression and positively adjust to age-related losses and declines. An interdisciplinary approach involving a physician, social worker, and occupational therapist is also useful to help adults and their families plan for this transition, especially when it is done early in the process rather than during a time of crisis. Recognizing that this is a process rather than a single event, and that there are individual differences in how older adults experience this process, will also ease the shift to other transportation options (Dobbs et al., 2002; Liddle et al., 2007). At the same time, there must be other mobility options available in the community that can meet the needs of individuals who are no longer able to drive.

Liddle, McKenna, and Broome (2004) recently reported results of a preliminary evaluation of the acceptability and effectiveness of resources developed to support the transition from driving in Queensland, Australia. Collectively, there was support for the development of a range of resources to meet the needs of individuals at different stages of the driving cessation process. Some of these resources provided an overview of information while others offered intensive support, information, and practical advice and skill development. Findings suggested that local information, peer involvement, and accessibility may be important issues in resource development. The authors noted that a wide range of resources are being developed and improved to meet the needs of older adults planning or experiencing driving cessation. Preliminary input from both older people (current and retired drivers) and health professionals led to several recommendations for improving awareness of and planning for driving cessation.

- Improve the media representation of retired drivers
- Provide attractive and acceptable transport alternatives to driving
- Increase awareness of driving cessation issues and successful transition strategies through education campaigns and awareness-raising talks and brochures
- Encourage planning for driving cessation by providing awareness-raising talks that discuss planning strategies and transport alternatives
- Link transport planning with other planning and health-care initiatives (e.g., link transport planning and retirement planning, provide education and training to health professionals to enable them to assist people with transport planning for the future

An approach that holds promise for helping older adults transition from driving has to do with how we can make our communities more livable. The issue of how livable our communities will be for us as we grow older is an important one and yet people rarely think about it until it has become clear that their needs are no longer being met. Now that our society is aging, the role of the physical and social environments in promoting independence and strengthening civic and social ties has become increasingly recognized.

A livable community has been defined as one that has affordable and appropriate housing, supportive community features and services, and adequate mobility options, which together facilitate personal independence and the engagement of residents in civic and social life (Kochera et al., 2005). One of the most important aspects of a livable community is the high level of engagement of its residents, ranging from participation in social activities and relationships, to volunteering, to civic participation in community planning and the political process. Such engagement is a vital part of successful aging and transportation is the means by which people not only connect to the goods and services, but also stay engaged. As part of the initiative to promote livable communities, two broad recommendations have been made in the area of transportation and mobility: (1) communities should facilitate driving by older adults by improving the travel environment, supporting driver education, and promoting safe driving throughout the lifespan; and (2) communities should take positive steps to enhance mobility options, including public transportation, walking, and bicycling, and specialized transportation for individuals with varied functional capabilities and preferences.

CONCLUSIONS

The transition from driving to nondriving can be a difficult and emotional time for older drivers, their families, and support networks. Efforts to smooth this process need to occur at many points along the "safe mobility" continuum that is anchored by safe independent driving at one end and dependence on other community mobility options at the other end. Unfortunately, there has been limited research on the factors that might lessen the adverse outcomes that can result from stopping driving. There is clearly a need to better understand the process of driving retirement among older adults and how it can be managed more effectively. In particular, the role of self-regulation in extending safe driving and easing the transition to nondriving is an area that would benefit from increased research attention. Self-regulation appears to hold promise for drivers without serious cognitive impairment as a behavioral adaptation for declining functional abilities. However, much remains to be understood about the extent to which older adults self-regulate, how self-regulation is associated with driving confidence and insight into functional impairments, and the extent to which self-regulation is warranted and appropriate. Also of interest is the type of self-regulation that is occurring, particularly at the higher levels of driver skills and controls. As discussed in Chapter 3, it is the decisions that older adults make at the "strategic" and "life goals" levels that will likely have the greatest potential to keep them safely mobile.

13 Other Community Mobility Options

*Birds have wings; they're free; they can fly where they want when they want.
They have the kind of mobility many people envy.*

Rodger Rory Peterson

INTRODUCTION

Older adults who are no longer able or choose not to drive must still be able to meet their transportation needs to retain their community mobility and hence their quality of life. It has been estimated that once individuals stop driving, most will be dependent on other community mobility options for several years—for men, about 6 years and for women, about 10 years (Foley et al., 2002). Unfortunately, few plan for the time when they will no longer be driving (Bailey, 2004; Kostyniuk & Shope, 2003). When the time comes, they often rely on friends and family to drive them, given their strong preference in the United States for the personal automobile (Kostyniuk & Shope, 1998). However, the availability and willingness of family and friends to provide rides have become increasingly constrained by trends toward smaller family size, higher divorce rates, and more women in the workplace (Federal Highway Administration, 1997). In addition, many older adults are reluctant to ask for rides, preferring to stay at home unless absolutely necessary (Freund, 1996).

Maintaining community mobility is further complicated by the increasing trend of people wanting to remain in their own homes. By aging in place (particularly in rural and suburban areas), older adults may have fewer transportation resources available to them than if they sought out more transportation-friendly retirement areas (Coughlin & Lacombe, 1997). Older adults living in rural areas face special transportation challenges because of the limited transportation services available and the long distances they must often travel to meet their health and social needs. In addition, older adults living in rural areas may be more vulnerable and isolated than their urban or suburban counterparts—they are more likely to be older (age 85 and older), poorer, and in worse health than those in urban and suburban areas (Kochera, et al., 2005).

A number of community mobility options have been developed to meet the mobility needs of older adults who no longer drive. Among these are traditional public transit (e.g., buses, light rail, trains, and subways); paratransit (demand response services including Americans with Disabilities Act [ADA] transit services) and specialized transit services (e.g., those operated by health and human service providers); private transit; supplemental transportation programs (e.g., operated by private-sector transit services, community groups, and volunteer groups); and other alternatives such as walking, bicycling, and using small motorized vehicles such as golf carts (Kerschner & Hardin, 2006; Suen & Sen, 2004). The extent to which these services are available varies from community to community. There is also considerable variation in how aware people are of them, how difficult they are to use, and how much they cost.

Research by the Beverly Foundation (2004b) has identified five attributes of community mobility options that determine whether they are "senior friendly," including availability, acceptability, accessibility, adaptability, and affordability. As described by the Beverly Foundation, accessibility has to do with whether people can get to and physically use the service. Acceptability has to do with how well the service meets the personal standards of users relative to such things as cleanliness of the vehicle, safety of the waiting area if there is one, and politeness of the driver. Adaptability has to do with whether the service is flexible enough to be responsive to the special needs of individual users, such as accommodating a person in a wheelchair or someone needing to make multiple stops on the same trip. Affordability has to do with whether the costs are within reach of users, and if there are options for reducing out-of-pocket costs through such things as discounts, vouchers, and coupons. This chapter describes various community mobility options, discusses the extent to which they are "senior friendly," and highlights efforts that have been undertaken (as well as recommendations) for making individual transportation services more available, acceptable, accessible, adaptable, and affordable for older adults, and for improving the transportation system overall.

EXISTING COMMUNITY MOBILITY OPTIONS

TRADITIONAL PUBLIC TRANSIT

Traditional public transit typically operates on a schedule with predetermined stops along a specified route, and can include buses, subways, light rail, or commuter rail (Suen & Sen, 2004). Fixed-route bus service is characterized by printed schedules or timetables, designated bus stops where passengers board and alight, and the use of larger vehicles (Alan M. Voorhees Transportation Center, 2005).

Overall, public transit accounts for a small percentage of trips made by older adults, with one study showing that only 1 percent of trips made by older adults were made by transit (Collia, Sharp, & Giesbrecht, 2003). Other research has found that about 12 percent of older adults overall report using it within the past year (Polzin & Chu, 2005). Older residents living in highly urbanized areas, however, are much more likely to use public transportation. For example, many older adults in the highest density areas take public transportation every day, contributing to an estimated 310 million total public transportation trips by older adults each year (Bailey, 2004). At the same time, public transportation is not available for much of the population—over a third of American households do not have public bus service within 2 miles of their homes, and in rural areas, over three-quarters of the population lack these services (Kerschner & Hardin, 2006; Kochera et al., 2005).

When public transportation is available in a community, there are often barriers to its use. Age, health, and disability all have strong impacts on the use of public transportation—one in three people over age 75 has a medical condition that restricts his or her ability to travel and one in six of these individuals reports that the condition limits use of public transportation (Kochera et al., 2005). Thus, it is often the case that the same deficits in abilities that caused an individual to have problems with driving also discourage the use of public transportation services. For example, older adults may have difficulty walking to the bus stop, waiting for the bus to arrive, climbing aboard, standing if no seats are available, and knowing when to get off at their stop. Other reasons reported for not using public transportation include safety concerns, lack of knowledge regarding use, and inconvenienc. (Burkhardt, McGavock, Nelson, & Mitchell, 2002; Coughlin, 2001; Kochera et al., 2005).

Older adults may be more likely to consider using public transportation options as improvements are made to better meet their needs. A number of recommendations for improving public transportation for older adults have been made and are listed below (see Burkhardt, 2003; Burkhardt, McGavock, & Nelson, 2002; Kerschner & Hardin, 2006).

- Improve schedule reliability
- Expand hours of operation
- Use advanced technologies to monitor vehicle locations and generate real time arrival and departure information
- Make it easier for older drivers to enter and exit the bus by reducing physical barriers such as steps, and operating low-floor vehicles
- Educate bus drivers to make them more responsive to passenger needs.
- Establish travel training for older adults not accustomed to public transportation
- Provide more user-friendly travel information (e.g., maps and schedules that are easy to read and understand)
- Focus on being responsive to customer needs and not just operating vehicles.
- Increase the number of reserved seats for older adults
- Make sure there are shelters and benches at all bus stops
- Involve older adults in transportation planning

Paratransit

Paratransit typically refers to demand-response transportation services (also called *dial-a-ride*), but also includes subscription bus services, shared-ride taxis, and car pooling and van pooling (Alan M. Voorhees Transportation Center, 2005). Paratransit is characterized by flexible routing and the use of relatively small vehicles that provide door-through-door, door-to-door, curb-to-curb, or point-to-point transportation (Bruff & Evans, 1999). It is more flexible than conventional fixed-route services but more structured than the use of personal automobiles, with individuals requesting services between certain locations at a certain time, usually requiring a reservation.

Suen and Sen (2004) described several levels of paratransit with increasing demands for driver assistance. In curb-to-curb service, drivers provide assistance from the curb in front of the trip origin to the curb of the destination, including helping passengers into and out of the vehicle and with wheelchairs. In door-to-door service, drivers provide assistance from the door of the building at the point of origin to the door or driveway of the destination, helping passengers from the door of the building to the door of the vehicle if necessary. In door-through-door service, drivers help passengers across the threshold of both the origin and destination buildings, including up and down steps and in and out of the vehicle.

Public transportation agencies are required by the ADA to provide paratransit services for individuals of all ages who cannot reach or use fixed-route buses because of a functional impairment (Bailey, 2004). Public transportation agencies can also contract with taxis to provide complementary paratransit to accommodate people with disabilities and in some cases, specialized transit services are available to provide door-to-door transportation in the form of vans operated by human-service and nonprofit agencies. However, the availability of these options is often limited because (1) they are available only where there are regular transit services (and thus not in rural areas); (2) even in urban areas, many do not live close enough to existing bus lines; and (3) most elderly are not eligible for specialized transit because their disability is not severe enough (Rosenbloom, 2003). In some smaller communities without regular bus service, paratransit may be available to the general public for a fee. However, these services generally require reservations and involve a shared ride, which is inconvenient or troublesome for many older adults (Kihl, Brennan, Gabhawala, List, & Mittal, 2005). Thus, although paratransit and other specialized services are essential to those who use them, they account for only a small percentage of trips made by older adults (Kochera et al., 2005).

Many of the recommended strategies for improving paratransit services are similar to those for improving public transportation (Burkhardt et al., 2002).

- Implement quality control measures such as complaint monitoring
- Provide driver training to improve passenger assistance
- Expand trip-making flexibility by increasing opportunities for multipurpose trips
- Provide customer service training to drivers and dispatchers to encourage friendly and responsive service
- Improve marketing and outreach efforts to older adults to make them aware of what services are available and how they can access them

PRIVATE TRANSIT

Private transit services such as taxis are also available in many communities but can be costly (Beverly Foundation, 2004a). When used as private transit, taxis can either be booked by telephone or hailed on the street, with the capacity to carry multiple passengers (Suen & Sen, 2004). Harbutt (2007) recently reviewed the pros and cons of taxi use by older adults. He found that taxis can be attractive to former drivers because of their similarity to the personal automobile, although the costs of taxis are often a barrier to their use (McKenzie & Steen, 2002). Taxis are also less attractive for a sizable number of older drivers who have trouble getting into and out of them or consider them to be unsafe (Stacey & Kendig, 1997).

Supplemental Transportation Programs

Supplemental transportation programs (STPs) are community-based transportation programs organized to meet the specialized mobility needs of older adults through trip chaining, transportation escorts, door-through-door service, and other means of personal support. They are intended to complement or supplement existing transportation services by reaching out to older adults (especially those age 85 and older) with special community mobility needs. Information has been collected on over 400 such programs since 2000 through an annual survey conducted by the Beverly Foundation (2004b). These programs vary considerably in terms of where they are located, how they are organized, ridership, trip purpose, use of escorts, type of vehicle, rider fees, drivers, and funding. However, survey findings indicate that the majority operate in either rural areas or a mix of rural/urban, are nonprofit, operate door-to-door service for older adults or individuals with disabilities, are used for medical purposes, operate during the daytime, employ paid and volunteer drivers, require either same-day or 24-hour notice, and are funded through grants or fees/donations from riders.

Common among STPs are volunteer ride programs that use private cars and other vehicles and are operated by private resources or volunteer drivers (Winter Park Health Foundation, 2006). Such programs may also be more affordable than public transportation, although they tend to have restricted hours and requirements for advanced scheduling. A successful model that uses both volunteer and paid drivers in private automobiles is the Independent Transport Network (ITN), a shared-cost program that provides door-through-door services (including help with carrying packages and other items) to adults age 65 and older and adults with visual impairment. ITN is membership based—users pay nominal dues, pay for their rides through personal transportation accounts (at roughly 50 percent of the cost), and accrue transportation credits in a variety of ways (see http://www.itnamerica.org for more information about ITN).

One benefit of volunteer driver programs is that they allow older adults to maintain their mobility without sacrificing their autonomy. However, an important barrier to the widespread adoption of such programs is the availability and affordability of liability insurance for drivers. Several strategies for maximizing the potential of volunteer-driver programs were recently identified by delegates to the 2005 White House Conference on Aging, in support of the resolution to "ensure older Americans have transportation options to retain their mobility and independence" (ranked as the third priority only behind long-term care and Medicare; White House Conference on Aging, 2005). Among the strategies were the following.

- Develop and fund policies that cover volunteer drivers for door-to-door and door-through-door transportation services, by local and state governments working with insurance companies
- Mandate insurance liability for volunteer drivers (no fault) to encourage volunteer programs

- Promote community-based volunteer transportation options and protect volunteer drivers from unreasonable insurance premiums
- Fund development of volunteer-based transportation for older adults including liability protection for volunteers

WALKING AND BICYCLING AND OTHER FORMS OF TRANSPORTATION

For older adults who are relatively physically fit, walking or bicycling may be a viable means of getting around, as well as a means of maintaining physical and functional health. However, walking is not without its risks (Harbutt, 2007). Older adults are at increased risk of death and serious injury as pedestrians (Oxley, Charlton, & Fildes, 2005). In addition, many older adults who are no longer able to drive because of declines in functional abilities are also unable to walk any distance for the same reasons. The frequency of walking among older adults in the United States is quite low—in one study, only 6 percent of adults age 65 and older made trips by foot, compared to about half of adults age 75 and older in Holland and Germany (Pucher & Dijkstra, 2003). Rates of walking in some European countries have been found to range from 30 to 50 percent (Whelan et al., 2006).

Bicycling is even more limited among older Americans and little has been done in the United States to address the need for a safe infrastructure for either walking or bicycling, including sidewalks, road crossings, and traffic signals for pedestrians, and bicycle lanes and road crossings for bicyclists. Without attention to these infrastructure issues, walking and bicycling will continue to hold risk for the older adult population, given their growing numbers in the population and their susceptibility to injury. Making communities bicycle friendly—that is, providing safe accommodation for cyclists and encouraging residents to bike for transportation and recreation—involves concerted efforts in a number of areas including engineering, education, encouragement, enforcement, and evaluation and planning (League of American Bicyclists, 2007).

One option that has gained in popularity among individuals who find walking or bicycling difficult or undesirable is the use of small motorized vehicles such as golf carts and scooters. Unfortunately, the safety of these forms of transportation is a concern (Whelan et al., 2006).

CREATING MORE SENIOR-FRIENDLY COMMUNITY MOBILITY OPTIONS

Recommendations for improving community mobility options overall were recently developed by Kerschner and Hardin (2006).

- Availability: make vehicles visible so people will see them; extend services to areas not previously served; focus service provision and marketing efforts in areas where older adults tend to congregate (e.g., senior centers, churches); provide transportation for life-enrichment activities as well as basic needs

- Acceptability: improve image of transit options (e.g., by linking with affinity transit services run by interfaith groups, volunteer groups, senior centers, or churches); involve older adults in planning, solicit their opinions, and take their preferences into account in providing and adapting services; establish principles for treatment of older adult passengers; ensure that communication and marketing efforts are personalized and dignify older adults; ensure the safety of stops and equipment
- Accessibility: provide training for drivers and staff to ensure they are respectful of and sensitive to older adult passengers; allow time for passengers to enter and exit the vehicle and consider flexible waiting rules for older adult passengers; use vehicles that are easily accessible by older adults
- Adaptability: provide flexible scheduling and destination choice; provide low- or no-cost options; broaden planning perspective to include idea of mobility rather than transportation; use vehicles that can be adapted to accommodate older adult preferences
- Affordability: consider creative approaches (e.g., ride free promotion, day pass allowing older adults to go wherever they want); find partners to share costs of services (e.g., shopping centers, physicians' offices, banks); seek discretionary funding to support supplemental services such as quality-of-life transportation services; provide opportunities for older adult passengers and volunteer drivers to make donations when fees are not required

TRANSPORTATION COORDINATION

Focusing on individual transportation services to ensure their responsiveness to the needs of older people is an important part of enhancing mobility. However, taken together, community transportation options are often fragmented and uncoordinated in communities. Therefore, it is also important to view individual transportation services within a given community as part of a system, and to determine where there may be gaps and where there may be opportunities for improved coordination and collaboration. Communities, working in concert with state and federal agencies, have an opportunity to forge transportation systems composed of different types of transportation services at different prices that best meet their unique community needs.

A number of federal initiatives over the past several years have increased the ability of states and communities to coordinate transportation services for older adults and those with disabilities. In 1990, the ADA was enacted to protect the civil rights of people with disabilities and ensure them access to employment, public transportation, public accommodations, and telecommunications. Since passage of the ADA, a wide range of federal initiatives including legislation, grant opportunities, and other programs have been undertaken to further increase mobility for older adults and people with disabilities.

In 2003, a memorandum of understanding was signed between the U.S. Administration on Aging and the Federal Transit Administration to assist their respective networks in the coordination of transportation services for older adults and to facilitate access to these services. The specific aims of the memorandum are to increase awareness about the transportation needs of older adults and strategies to address these needs; establish baseline data on older driver transportation services; develop and implement a joint plan to provide ongoing technical assistance and training to state and local organizations to promote promising practices for coordination; work collaboratively with stakeholders at the federal, state, and local level to identify barriers and solutions to accessing transportation services; and work together to better coordinate the provision of funding opportunities to foster coordination of transportation services and the development of innovative service delivery models.

In 2004, The Federal Interagency Coordinating Council on Access and Mobility (CCAM) was established through Executive Order 13330 entitled "Human Service Transportation Coordination" to coordinate 64 federal programs providing transportation funding for older Americans, people with disabilities, and individuals with low income. Composed of 11 departments and agencies (U.S. Departments of Transportation, Health and Human Services, Labor, Education, Agriculture, Housing and Urban Affairs, Interior, Veterans Affairs, Social Security, the Attorney General's Office, and the National Council on Disabilities), CCAM is responsible for seeking ways to reduce transportation service duplication, increase efficient transportation service delivery, and expand transportation access for older Americans, people with disabilities, and individuals with low income.

To implement the executive order, CCAM launched United We Ride, an interagency initiative to coordinate human service transportation across all levels of government and the private and nonprofit sectors. The specific aims of United We Ride are to develop an education plan for coordinated human service transportation to enhance customer access at the local level; simplify access to transportation services and enhance customer service through the development of a comprehensive and coordinated transportation system; reduce restrictive and duplicative laws, regulations, and programs related to human service transportation at the federal level; ensure comprehensive planning for the coordination of human service transportation for individuals with disabilities, older adults, and individuals with lower income; standardize cost allocation processes; and document successful strategies in coordinating human service transportation at the federal, state, and local governments.

The Safe, Accountable, Flexible, and Efficient Transportation Equity Act—A Legacy for Users Public Law (SAFETEA-LU) was enacted in August of 2005 to reduce barriers to transportation and provide services beyond ADA requirements. It supports federal transit programs and establishes an upper limit on the amount of funds that can be made available. In 2006, the Older American Act was amended (Title IV Activities for Health, Independence, and Longevity) to incorporate provisions aimed at improving transportation by including planning for baby boomers, assessing technology-based models, improving support for family caregivers, and building awareness of cognitive impairments such as dementia.

Strategies found to be effective for improving transportation coordination include (U.S. Department of Health and Human Services, 2005).

- Establishing broad-based coalitions and partnerships
- Coordinating planning through ongoing relationships with local planning and development agencies
- Leveraging funding from a variety of sources

- Paying careful attention to the specific objectives and regulations of federal transportation programs, given that much of the funding originates with federal programs aimed at the unique needs of individual populations
- Integrating new technologies into operations to improve efficiency and responsiveness to users

CONCLUSIONS

Each of the community mobility options described in this section carries with it strengths and weaknesses in terms of being available, acceptable, accessible, adaptable, and affordable for older adults who are unable or choose not to drive. Improving the senior friendliness of individual services can go a long way toward enhancing the mobility of older adults. However, it is clear that no single community mobility option will be able to meet the needs of its older adult population, given the diverse needs and preferences of older individuals. Successful efforts to enhance community mobility will require a wide range of public and private services and systems, increasingly referred to as a "family of services" (U.S. Department of Health and Human Services, 2005). It is also clear that increasing the use of community mobility options by older adults will likely require strong promotion, with information being presented in a form that is readily comprehensible, especially given the lack of experience that older adults have with many alternative transportation options (Whelan et al., 2006). One approach that holds promise for helping to accomplish this is the use of personal community mobility managers—individuals who serve as one-stop resources for older drivers (e.g., through a telephone hotline) to provide information on all aspects of maintaining community mobility. To be effective, this approach must go hand in hand with a broad-based, coordinated transportation system that is fully integrated into the community.

Appendix: Driver Evaluation Tools

I still have nightmares about taking tests.

Bruce Jenner

INTRODUCTION

Driving is a complex task requiring an integration of visual, cognitive, and psychomotor skills in order to safely manage the driving environment. As such, the task of evaluating these skills is an equally complicated process. Further complicating this process is the lack of empirical data linking evaluation tool outcomes to driving performance or crash risk. While research in this area is being conducted, there is currently a paucity of studies. It is beyond the scope of this book to review the strengths and weaknesses of each available driver evaluation tool. Instead we provide here a brief description of the more commonly used evaluation tools. This appendix is divided into three main categories: perceptual, cognitive, and psychomotor evaluation tools.

PERCEPTUAL EVALUATION TOOLS

Dynamic Visual Acuity

Structure from Motion

Structure from motion (SFM) is the perceptual ability to perceive three-dimensional (3-D) shape from motion parallax information. SFM is studied by simulating animated 3-D objects constructed of points of light on a computer screen. Thresholds for detecting SFM are usually determined by adding "noise" to the display (i.e., points of light that are moving independent of the simulated object points) and determining the signal-to-noise level where the object can be accurately identified 75 percent of the time. Some research has attempted to relate SFM performance to driving performance outcome measures (Rizzo, Reinach, McGehee, & Dawson, 1997; Uc, Rizzo, Anderson, Shi, & Dawson, 2005; Uc, Rizzo, Anderson, Shi, & Dawson, 2006). Collectively this research has found that SFM scores were predictors of at-fault safety errors during on-road driving; there was no relationship between SFM scores and rear-end crashes, abrupt slowing, or premature stopping in a simulator study; and reduced 3-D SFM ability was a strong predictor of crashes in a driving simulator.

Moving "E" Test

This test involves the ability to perceive details when there is relative motion between the person and stimuli. So that motion thresholds can be accurately measured, computer-generated moving stimuli are generally used. One study found a weak but significant correlation between scores on dynamic visual acuity and incidents of unsafe driving on an on-road driving assessment (McKnight & McKnight, 1999).

Static Visual Acuity

Static visual acuity can be measured with a number of different charts, all of which can ultimately express visual acuity as the log minimum angle resolvable, and usually is expressed on a fractional scale (20/10–20/600). A large number of studies have related acuity scores to driving outcomes. Overwhelmingly, studies find a poor relationship with acuity scores to any driving outcomes measure when corrected vision is better than about 20/70 (Decina & Staplin, 1993; Guerrier, Manivannan, & Nair, 1999; Janke, 2001; Kantor et al., 2004; Laux & Brelsford, 1990; Marottoli et al., 1994; McCarthy & Mann, 2006; McCloskey et al., 1994; Owsley, McGwin et al., 1998; Owsley et al., 2002; Owsley et al., 1999; Owsley et al., 2001; Ranney & Pulling, 1990; Sims et al., 1998; Szlyk et al., 1992; Szlyk et al., 1993; Szlyk et al., 2005; Szlyk et al., 2004; Szlyk et al., 1995; Uc et al., 2005; Uc et al., 2006).

Contrast Sensitivity

Contrast sensitivity can be measured with a number of different charts and methods, all of which can ultimately be expressed as log minimum contrast sensitivity (0.5–2.0). There is good evidence that poorer contrast sensitivity scores are related to poorer outcomes on driving measures, particularly when contrast sensitivity is equal to or less than 1.25 (Ball et al., 1993; Bowers, Peli, Elgin, McGwin, & Owsley, 2005; Hennessy, 1995; Ivers, Mitchell, & Cumming, 1999; Janke, 2001; Janke & Eberhard, 1998; Kantor et al., 2004; Margolis, Kerani, McGovern, Songer, Cauley, & Ensrud, 2002; McKnight & McKnight, 1999; Owsley et al., 1991; Owsley et al., 1998; Owsley et al., 2002; Owsley et al., 1999; Owsley et al., 2001; Sims et al., 1998; Szlyk et al., 2005; Szlyk et al., 2004; Szlyk et al. 1995; Uc et al., 2005; Wood, 2002; Wood & Troutbeck, 1995).

Visual Field

The visual field refers to the range of what a person can see without moving his or her eyes. The visual field is usually discussed in terms of the central field, where visual acuity is greatest, and the peripheral visual field. In general, the visual field is measured by having a person fixate his or her gaze on a point and indicate when he or she sees stimuli appearing at various places in the visual field. Performance is usually measured in terms of the number of degrees from central (0 deg) where stimuli can be perceived. Loss of vision in the central visual field is also possible, as is the case for people with macular degeneration and other ophthalmologic diseases.

There is conflicting evidence that people with central visual field loss are at higher risk of crashes and violations and perform more poorly on driving simulator tasks (Ball et al., 1993; Bowers et al., 2005; Coeckelbergh et al., 2002; Decina & Staplin, 1993; Ivers, Mitchell, & Cumming, 1999; Janke & Eberhard, 1998; Kantor et al., 2004; Klavora et al., 2000; Owsley et al., 1991; Owsley, McGwin et al., 1998; Owsley et al., 1999; Schanke & Sundet, 2000; Szlyk et al., 1992; Szlyk et al., 1993; Szlyk et al., 2005; Szlyk et al., 2004; Szlyk et al., 1995; Szlyk et al., 1995; Wood & Troutbeck, 1995).

COLOR VISION/DISCRIMINATION

Color vision/discrimination is the ability to perceptually distinguish among different wavelengths of light. There are numerous tests for this ability. As discussed by Owsley and McGwin (1999), several states test for color discrimination abilities during license renewal, presumably to assess whether or not a person can distinguish the colors of a traffic control signal. Even in a person who is color deficient, there is other information that can be used for perceiving the state of a traffic signal, including relative location and luminance. No study has shown that color deficiency results in compromised driving ability or increased risk of a crash (Ball et al., 1993; Owsley et al., 1991; Owsley & McGwin, 1999; Vingrys & Cole, 1988).

STEREOACUITY

Stereoacuity refers to the perceptual ability to resolve the slightly different images reaching each eye that result from differing depths of parts of objects in the visual scene, allowing the brain to perceive depth. This ability is measured by a number of devices, all of which display different images to each eye. The differences in the images are calibrated so that stereoacuity can be measured by the least amount of difference that can be perceived as a depth difference. Scores range from 15–600 seconds of arc, with the higher end indicating stereoblindness. No study has shown that poor stereoacuity results in compromised driving ability or increased risk of a crash (Margolis et al., 2002; Owsley et al., 1991; Kantor et al., 2004; Owsley, McGwin et al., 1998).

VISUAL ATTENTION

Useful Field of View

The useful field of view (UFOV®) test, developed by Ball and colleagues, assesses the size of the visual field over which people can respond to rapidly presented information. This test relies on both perceptual and cognitive abilities, such as selective and divided attention. Ball and colleagues have conducted extensive research relating scores of the UFOV test and various measures related to high-risk driving (Ball et al., 1993; Ball et al., 2005; Bower et al., 2005; Duchek, Hunt, Ball, Buckles, & Morris, 1998; Goode, Ball, Sloane, Roenker, Roth, Myers et al., 1998; Myers, Ball, Kalina, Roth, & Goode, 2000; Owsley et al., 1998; Owsley et al., 1991; Owsley,

McGwin et al., 1998; Sims et al., 1998). Collectively, these studies have found that people with reductions in the UFOV perform more poorly on on-road driving tests, have a history of at-fault crashes, and are more likely to be in future at-fault crashes. One study found that drivers with a 40 percent or greater reduction in the UFOV were 2.2 times more likely to be in a crash during a 3-year follow up. Another prospective cohort study with a large sample of drivers found that drivers who took 353 ms or longer on subtest 2 of the UFOV test were 2.02 times more likely to be in an at-fault crash in the following 4–5 years.

Number Cancellation Task

The test involves marking out all the numbers in a row that match a circled number at the far left-hand side of a row within a given amount of time. Studies have found that poorer performance on this task is significantly correlated with self-reported crashes and violations, as well as problems found during on-road driving evaluation. Richardson and Marottoli (2003) found a correlation of .43 between the scores on the number cancellation task and an on-road driving assessment.

Hearing

There are numerous tests for hearing sensitivity, both objective and subjective. Hearing sensitivity has been measured in many studies addressing the relationship between cognitive and perceptual measures to measures of at-risk driving (Sims et al., 1998; McCloskey et al., 1994; Marottoli et al., 1994; Szlyk et al., 2002; Sims et al., 2000; Mercier, Mercier, O'Boyle, & Strahan, 1997; Kantor et al., 2004). Most of these studies do not report the relationship between hearing ability and unsafe driving presumably because there was no significant relationship found. In those studies that have reported hearing outcomes, no significant relationship was found between hearing loss and police-reported injury crashes or between hearing loss and simulated driving performance.

COGNITIVE EVALUATION TOOLS

Visuospatial Abilities

Hooper Visual Organization Test

The Hooper Visual Organization Test (HVOT) measures an individual's ability to organize visual stimuli. The test takes approximately 15 minutes to administer and consists of 30 line drawings, each showing a common object (such as an apple or a ball) that has been cut into several pieces. The pieces are scattered on the page like parts of a puzzle, and the individual's task is to tell what the object would be if the pieces were put back together correctly. The HVOT has been utilized in several studies to understand the connection of visuospatial ability to driving (Marottoli et al., 1998; Richardson & Marottoli, 2003). Results of borderline significance ($p = 0.059$) have been found between the HVOT and a self-reported history of crashes, moving violations, or being stopped by police; however, during that study the relationship did not hold during multivariable analysis. The HVOT has also been compared to on-road driving but no significant correlations were found.

Embedded Figures Test

The Embedded Figures Test (EFT) is used to measure perceptual style and analytical ability. The test requires finding simple forms that are embedded in larger figures. The score is the average time in seconds to detect the forms. As such, higher scores reflect greater difficulty in analyzing a part separately from a wider pattern. Sometimes the number correct is used instead of average time. The EFT has been used widely and compared to various measures of driving (Demick & Harkins, 1999; Guerrier, Manivannan, & Nair, 1999; James et al., 1997; Marottoli et al., 1998; Mercier et al., 1997; Ranney & Pulling, 1990). Studies have found EFT scores to be a better predictor of driving than age. EFT scores have also been found to be predictive of driving problems, and several studies have found it to be significantly correlated with age. In other studies, scores on the EFT were not significantly correlated to self-reported history of crashes, moving violations, or being stopped by police.

Rey-Osterreith Complex Figures Test

The Rey-Osterrieth Complex Figure Test (RO-CFT) is used for the evaluation of visuospatial constructional ability and visual memory. It consists of three test conditions: Copy, Immediate Recall, and Delayed Recall. During the first step, participants are given the RO-CFT stimulus card and then asked to draw the same figure. Subsequently, they are instructed to draw what they remembered. Then, after a delay of 30 minutes, they are required to draw the same figure once again. Each condition of the RO-CFT takes 10 minutes to complete and the overall time of completion is about 30 minutes. Scores on the RO-CFT have been related to driving performance measures in several studies (Ball et al., 1993; Galski, Bruno, & Ehle, 1992; Goode, et al., 1998; Lundberg et al., 1998; Rizzo et al., 2001; Uc et al., 2005; Uc et al., 2006). This work has found scores to be significantly correlated with crash avoidance, evasive action, and threat recognition actions during driving simulation activities. RO-CFT scores were significantly predictive of crashing or unsafe actions during the driving scenarios. Scores on the test have also been found to be related to on-road driving performance in general and the ability to properly identify landmarks.

Block Design Test

The Block Design Test (BDT) is a subtest on many intelligence tests that assesses visuospatial and motor skills. The person taking the test is required to arrange blocks that have all white sides, all red sides, and red and white sides according to a specific pattern. People are timed on this task and compared to a normative sample. Results from the BDT have been compared directly to on-road driving and simulator driving (Ball et al., 1993; Galski et al., 1992; Lundberg et al., 1998; Rinalducci et al., 2001; Rizzo et al., 2001; Schanke, & Sundet, 2000; Sims et al., 1998; Szlyk et al., 2002; Uc et al., 2005; Uc et al., 2006). Collectively these studies have found significant correlations between the BDT and actual and simulated driving; correlations between participants' ability to properly identify landmarks and BDT scores; significant correlations with obeying speed limits, crossing the median, crashes, leaving the road, abrupt slowing, and age; and significant correlations with crash outcomes.

Letter Cancellation Test

The Letter Cancellation Test (LCT) consists of six 52-character rows in which the two target characters are randomly interspersed approximately 18 times in each row. Participants are asked to cross out the target characters as quickly as possible until they finish all the rows. LCT scores have been compared to both on-road and simulated driving (Galski, Bruno, & Ehle, 1992; Whelihan, DiCarlo, & Paul, 2005). A significant correlation was found between actual driving and LCT in one study, but no significant relationship was found in another. During simulated driving, LCT scores were related to both braking and steering behavior.

Benton Judgment of Line Orientation

The Benton Judgment of Line Orientation (JLO) test is widely used to assess visuospatial processing. The test requires the participant to identify which two of 11 lines presented in a semicircular array have the same orientation in two-dimensional space as two target lines. Scores on the JLO have been compared to driving simulation, on-road driving, and landmark identification (Szlyk et al., 1995; Uc et al., 2005; Uc et al., 2006). During on-road performance, JLO scores were correlated with better backing-up scores, and one's ability to properly identify landmarks. In a simulator study, scores on the JLO were found to be significantly different between a group with Alzheimer's disease and normal controls. The scores were also found to be related to "crash or risky avoidance behavior" and "abrupt slowing," but not to "premature stopping." When adjusting for groups, only the "abrupt slowing" finding was significant.

Wisconsin Card Sorting Test

During the Wisconsin Card Sorting Test (WCST) a number of stimulus cards are presented to the participant. He or she is then asked to match each card in an additional stack to one of the stimulus cards, thereby forming separate piles of cards for each. The participant is not told how to match the cards; however, the participant is told whether a particular match is right or wrong. The test takes approximately 12–20 minutes to complete. While several studies have utilized the WCST as part of a battery of tests to be compared to driving measures, significant correlations between WCST scores and simulated and on-road driving measures have not been found (Lundqvist, Gerdle, & Roennberg, 2000; Whelihan et al., 2005).

Maze Navigation Test

The Maze Navigation Test (MNT) consists of a set of paper forms on which the participant is required to trace a path through a drawn maze of varying complexity. There is no time limit for this test. The participant must avoid blind alleys and dead ends; no backtracking is allowed. Two studies have found significant relationships between the MNT and on-road evaluations (Galski et al., 1992; Whelihan, DiCarlo et al., 2005). In addition, correlations between MNT scores and both proper braking and steering behavior were found during a simulator study.

Visual Form Discrimination Test

The Visual Form Discrimination (VFD) test is a 16-item multiple-choice test requiring visual analysis and matching sets of geometric shapes. It is typically used to assess the ability to make fine visual discriminations. VFD scores have been compared to both on-road and simulated driving (Szlyk et al., 2002; Whelihan et al., 2005). During a simulator study, the VFD was correlated with horizontal eye movement. In addition, VFD scores were found to be significantly different between groups divided in terms of scores on the mini mental status exam. However, in a different study utilizing on-road driving, no significant relationship was found between VFD scores and driving.

Motor Free Visual Perception Test

During the Motor Free Visual Perception Test (MVPT), participants are given stimuli depicting four incomplete figures and one whole figure. Participants select the incomplete figure that, when completed, would match the target figure. The number of errors is recorded as the outcome. This measure has been compared to crash records, on-road driving, and self-reported driving history (Ball et al., 2006; Kantor et al., 2004; Lesikar, Gallo, Rebok, & Keyl, 2002). Utilizing crash records, the MVPT was found to be significantly predictive of at-fault crashes (even when controlling for age, gender, and miles driven). MVPT scores between the group of at-fault crash-involved participants and the controls were found to be significantly different. The MVPT was also utilized in a study that included an on-road evaluation, but the MVPT was not included in the final statistical model, suggesting that it was not significantly related. When considering MVPT scores and self-reported driving history, one study shows large estimates of relative risk (2.83 times as likely to report a crash), but the 95 percent confidence intervals for the point estimates of the relative risk included the null.

EXECUTIVE FUNCTION

Clock Drawing Test

The Clock Drawing Test (CDT) evaluates cognitive function including comprehension, memory, visuospatial abilities, abstract thinking, and executive function. It generally takes less than 5 minutes to administer. There are many variations to administering the CDT, as well as different scoring criteria. In one version of the CDT, subjects are verbally instructed to draw a clock, write all the numbers in, and set the time at 10 minutes after 11. The instructions are also written at the top of the page in 16-point font. There have been mixed results when examining studies that have compared the clock drawing test to several driving outcome measures (Fitten et al., 1995; Freund, Gravenstein et al., 2005; McCarthy & Mann, 2006; Ott, Heindel, Whelihan, Caron, Piatt & DiCarlo, 2003). Collectively these studies have found that CDT scores were a significant predictor of on-road driving performance; had a high level of accuracy in predicting driving outcomes in a simulator; and did not relate significantly to caregiver-reported driving problems.

Controlled Oral Word Association Test

The Controlled Oral Word Association Test (COWAT) is a measure of a person's ability to make verbal associations to specified letters (i.e., C, F, and L). This test is typically used for assessing verbal fluency and the ease with which a person can think of words that begin with a specific letter. It is able to detect changes in word association fluency often found with various disorders. Scores on the COWAT have been compared with on-road driving and simulated driving (Ott, Heindel, Whelihan, Caron, Piatt, & DiCarlo, 2003; Rizzo, McGehee, Dawson, & Anderson, 2001; Uc, Rizzo, Anderson, Shi, & Dawson, 2005; Uc, Rizzo, Anderson, Shi, & Dawson, 2006; Whelihan, DiCarlo, & Paul, 2005). Results of the on-road studies have found a significant, but weak relationship between COWAT scores and landmark identification, but no significant correlation between the COWAT and driving evaluation. Other work with driving simulators has found that COWAT scores were significantly predictive of crashing.

Trail Making Test (Parts A and B)

The Trail Making Test—Part A (TMT-A) tests the examinee's ability to connect circles numbered 1 to 25, scattered across a page of paper, in sequence in a specific length of time. The Trail Making Test—Part B (TMT-B) tests the examinee's ability to connect circles containing the letters A through L, and 13 numbered circles intermixed and randomly arranged. Individuals are instructed to connect the circles by drawing lines alternating between numbers and letters in sequential order, until they reach the circle labeled "End." If individuals make mistakes, the mistakes are quickly brought to their attention, and they continue from the last correct circle. The test takes approximately 5 to 10 minutes to complete (including parts A and B). Part A is generally used as a warm up for part B and has not been found to be correlated to driving outcomes. The TMT-B has been utilized extensively and compared with several different driving measures, including crash records, on-road driving, simulated driving, and self-reported incidents (Ball et al., 2006; Cushman & Cogliandro, 1997; Goode et al., 1998; Kantor et al., 2004; Lundqvist et al., 2000; Margolis et al., 2002; Marottoli et al., 1998; McCarthy & Mann, 2006; Ott et al., 2003; Rinalducci, Mouloua, & Smither, 2001; Rizzo et al., 2001; Schanke & Sundet, 2000; Stutts, Stewart, & Martell, 1998; Szlyk et al., 2002; Uc et al., 2005; Uc et al., 2006; Whelihan et al., 2005). Collectively this research has shown that TMT-B scores are significantly predictive of at-fault crashes; associated with crash risk, even after controlling for age, race, and measures of driving exposure; significantly associated with on-road driving ability; significantly related to participants' ability to identify specific landmarks during an on-road driving task; and significantly correlated with self-reported history of crashes, moving violations, or being stopped by police.

Benton Visual Retention Test

The Benton Visual Retention Test (BVRT) assesses visual perception, visual memory, and visuoconstructive abilities. It has three alternate forms, each of which consists of ten designs. Generally, the participant is shown ten designs, one at a time, and asked to reproduce each one on paper, as accurately as possible from memory. This measure has been utilized in conjunction with both on-road and simulated driving (Galski et al., 1992; Rizzo et al., 2001; Uc et al., 2005; Uc et al., 2006). When compared to simulated driving, BVRT scores were significantly predictive of crashing and were found to be significantly predictive of "crash or risky avoidance behavior" and "abrupt slowing," but not "premature stopping." BVRT scores were significantly correlated with participants' scores on a landmark identification driving task and with at-fault safety errors.

The Stroop Neuropsychological Screening Test

The Stroop Neuropsychological Screening Test (SNST) consists of two parts. In the Color Task, the individual reads aloud a list of 112 color names in which no name is printed in its matching color. In the Color-Word Task, the individual names the color of ink in which the color names are printed. This measure was used in a study that included a 2-hour on-road driving test with a driving instructor (Schanke & Sundet, 2000). During that study, SNST scores were not found to be significantly different between a group identified as "not suited for driving" and the "suitable" group. In a different study, the SNST was used as part of a larger battery, along with on-road and protected driving, but results of that specific test were not discussed in relation to any driving outcomes (Demick & Harkins, 1999).

DIVIDED ATTENTION

Simultaneous Capacity Test

Divided attention can be assessed using the Simultaneous Capacity Test. There are various forms of administering this test, but generally it consists of a background task and a foreground task. In one form of the test, the background task has random digits appear consecutively in the center of the screen at a pace of 10 seconds. The task is to scan for three consecutive odd digits, responding within 2 seconds with the nondominant hand. The foreground task consists of short messages appearing at random frequent intervals in the upper part of the screen. Many of these can be ignored, since they require no response, but other messages (such as "Press Key 9") require a response with the dominant hand within 2 sseconds. Percent correct responses, and three types of errors (omission errors, commission errors, and wrong responses) are tallied. The test continues for 10 minutes with increasing difficulty. This test was not able to predict the outcome variables of driving performance in a simulator or on the road (Lundqvist et al., 2000; Lundberg et al., 1998; Myers, Ball, Kalina, Roth, & Goode, 2000).

SELECTIVE ATTENTION

Zimmermann/Fimm's Incompatibility Test

Mental flexibility, the ability to switch from automatic to controlled processes, can be measured using Zimmermann/Fimm's Incompatibility test. It is a computer-controlled task that involves suppressing one's automatic response and implementing a controlled process. Compatible and incompatible items are presented in random order. The standard deviation of the reaction time to the correct items reflects the reaction speed variability due to the alternation between compatible and incompatible items. This test did not reveal a significant correlation between the neuropsychological test score and road test (DeRaedt & Ponjaert-Kristoffersen, 2000).

Dot Counting Task

The dot counting task is used to assess selective attention with visual search. While there are several different procedures for administering the test, the overall goal of the test is the same. A random pattern of dots is displayed in the central field of view on a computer screen. In some administrations, the background is dynamic, such as a moving road beneath the dots, while others have a solid, unmoving background. Subjects are asked to count the number of dots appearing on the screen. Mean search time and number of errors determine the test score. Research shows that scores on the dot counting task are correlated with the on-road test score, but are not good at identifying at-risk drivers or predicting crashes (Coeckelbergh et al., 2002; DeRaedt & Ponjaert-Kristoffersen, 2000).

SUSTAINED ATTENTION

Adaptation of Mackworth Clock Test

The Mackworth Clock test is a 45-minute test for measuring vigilance. It measures readiness to react during prolonged visual search, and gives insight into performance requiring arousal and sustained attention. It is sensitive to drowsiness in a low arousal state, and it demonstrates deterioration of continuous performance over time. The task provides no feedback to the subject and uses an infrequent signal in a noisy background. An adaptation of the Mackworth Clock test was used to evaluate subjects' performance of a prolonged, unvarying task, which is relevant to monotonous highway driving conditions (Fitten et al., 1995). The outcome was measured against the drive score from the Sepulveda road test, which consisted of a six-stage driving course 2.7 miles long. Driver score was found to be moderately correlated with this task.

WORKING MEMORY

Wechsler Memory Scale—Visual Reproduction

Visual reproduction is a subset of the Wechsler Memory Scale, which measures immediate and delayed visual memory. It requires memorization of a visual stimulus and construction of this stimulus from memory. Participants view a stimulus line design for 10 seconds and then are asked to reproduce the design from memory. The number of stimulus cards used varies by study, but all drawings are scored according to standard criteria. Several studies have found the test to be significantly related to crash status, driving simulator performance, on-road test performance, and specific driving maneuvers such as responding to vehicles/pedestrians, granting right-of-way, and positioning car for turn (Goode et al., 1998; Szlyk et al., 2002; Lesikar, et al., 2002; Marottoli et al., 1998).

Wechsler Adult Intelligence Scale—Digit Symbol

The Digit Symbol test is a test of speed of information processing that involves sub-stituting a symbol for a random succession of numbers. The subject is given a piece of paper with nine symbols corresponding with nine digits. Next on this piece of paper are three rows of digits with empty spaces below them. The subject is asked to fill in as many corresponding symbols as possible in 90 seconds. A figure is scored correct if it is clearly identifiable as the keyed figure, even if it is drawn imperfectly or if it is a spontaneous correction of an incorrect figure. Many studies have found no correlation between scores on the digit symbol test and driving outcome; however, one found a weak but significant correlation with lane boundary crossing within a simulated drive test (Lundqvist et al., 2000; Margolis et al., 2002); Rinalducci, Mouloua, & Smither, 2001; Marottoli et al., 1998; Szlyk et al., 2002).

Wechsler Adult Intelligence Scale—Digit Span

The Digit Span Subtest measures attention and mental manipulation. Digits are pre-sented consecutively on the screen, with an interval of 1 second. The participant has to enter the digit series in the same order on the keyboard, and then press a "finished" key. The number of digits in each item is process controlled, starting with three dig-its. After 13 forward items, 11 items follow, where the order of digits is to be reversed in the response, starting with two digits. The median of the number of recalled digits of the three last items in each task (forward and backward) is recorded. Several stud-ies have found conflicting evidence regarding significant correlations between the Digit Span Test and various measures of driving performance, such as brake pedal pressure, horizontal eye movement, and turning the wrong way into a one-way situ-ation (Laux & Brelsford, 1990; Schanke & Sundet, 2000; Sims et al., 1998; Szlyk et al., 2002).

Short Blessed Test

The Short Blessed Test is a measure of memory, orientation, and concentration, and is used to screen for possible cognitive impairment or dementia. Scores range from 0 to 28, with lower scores indicating better performance. This six-item test has been widely used in community epidemiological studies to indicate the probable presence of dementia. It has been found to be sensitive to both the presence and severity of dementia. The test requires identifying the current year and month, identifying time within 1 hour, counting backwards from 20 to 1, saying the months in reverse order, and repeating a name and address that the test administrator has told the subject earlier in the questioning. Many studies find no significant correlation between test scores and driving performance measures (Cushman & Cogliandro, 1997; Hunt et al., 1997; Stutts et al., 1998; Trobe, Waller, Cook-Flannagan, Teshima, & Bieliauskas, 1996).

Word Recall Memory Test

The Word Recall Memory test is administered in a variety of ways, with the number of words presented generally varying. After 12–20 monosyllabic words have been read to a person, he or she was asked to recall as many words as possible, and several minutes later asked again to recall as many of the words as possible. In the absence of clinically relevant cutpoints, poor performance is based on the lowest tertile of scores. Research has related scores on this test to driving outcomes (Foley et al., 1995; Lundberg et al., 1998) and has found that scores are related to increase in the risk of crashing, and suspended drivers with crashes had a significantly lower score on this test than did suspended drivers without crashes.

Mini-Mental Status Examination

The Mini Mental Status Examination (MMSE) is an 11-question measure that tests six areas of cognitive function: orientation, registration, attention and calculation, recall, language, and visuospatial perception/praxis. The maximum score is 30. Generally scores of 28–30 are considered normal, 21–27 suggest mild dementia, 11–20 moderate dementia, and 0–10 severe dementia. The MMSE takes only 5–10 minutes to administer. The MMSE has been used in numerous studies to understand the relationship between cognitive function and driving ability (Cox, Quillian, Thorndike et al., 1998; Fitten et al., 1995; Fox et al., 1997; Janke, 2001; Johansson, Bronge, Lundberg, Persson, Seideman, & Viitanen, 1996; Kantor et al., 2004; Lesikar et al., 2002; MacGregor, Freeman, & Zhang, 2001; Margolis et al., 2002; Marottoli et al., 1994; Marottoli et al., 1998; Odenheimer et al., 1994; Ott et al., 2003; Rinalducci, Mouloua, & Smither, 2001; Szlyk et al., 2002; Trobe et al., 1996; Whelihan, DiCarlo, & Paul, 2005). Collectively this research has found significant correlations between MMSE score and drive score; that the specificity and sensitivity of the MMSE was not sufficient for effective prediction of driving outcome measures; and that those with borderline cognitive impairment were more likely to have adverse events than those with higher or lower MMSE scores.

Clinical Dementia Rating

The Clinical Dementia Rating (CDR) is a numeric scale used to quantify the severity of symptoms of dementia. The CDR is based on a 90-minute interview with the subject and a collateral source. This interview assesses the subject's cognitive abilities in the areas of memory, orientation, judgment and problem solving, values of affairs, home and hobbies, and personal care. According to this scale, CDR = 0, .5, 1, 2, and 3 represent no dementia, very mild dementia, mild dementia, moderate dementia, and severe dementia, respectively. The CDR has been widely used to assess how severity of dementia relates to driving performance (Duchek et al., 2003; Hunt et al., 1997; Johansson et al., 1996; Ott et al., 2003). These studies have found significant correlations between CRD and simulated driving performance; that as dementia severity increases, driving performance scores decrease; and that scores are significantly correlated with self-reported crashes and driver history records of citations.

Mattis Organic Mental Syndrome Screening Examination

The Mattis Organic Mental Syndrome Screening Examination (MOMSSE) is a brief neuropsychological screening instrument specifically designed to assess cognitive status in the elderly. It consists of several Wechsler Adult Intelligence Scale subtests, a Benton geometric figure, and items from the Eisenson Test of Aphasia. The

MOMSSE test takes about 15–20 minutes to administer and evaluates the following 13 categories of cognitive function: information, abstraction, digit span, orientation, verbal memory, visual memory, speech, naming comprehension, sentence repetition, writing, reading, drawing, and block design. Scores range from 0 to 28, with lower scores representing higher functioning. A score greater than 9 represents cognitive impairment. Several studies have employed the MOMSSE as a way to compare cognitive impairment to measures of driving performance (Ball et al., 1993; Goode et al., 1998; Sims et al., 1998; Owsley et al., 1991; Owsley, McGwin, et al., 1998; Owsley et al., 2001; Sims et al., 2000). These studies report that MOMSSE scores correlated significantly with crash frequency, but did not adequately identify drivers at risk for crash involvement; and MOMSSE scores were not a significant predictor of at-fault crashes or distinguishing crashers from noncrashers.

Traffic Sign Recognition Test

There are many different variations of the Traffic Sign Recognition Test (TSRT). Because traffic signs are different across the nation, it is important to create a test that is relevant to the area in which the study is taking place. Therefore, many states have a different form of the TSRT. Generally, the test consists of a variety of traffic signs that are presented to a subject. The subject must identify and explain the meaning of each sign. There are also different scoring criteria for the various tests. Studies relating TSRT scores to driving outcomes have found that these scores correlated well with on-road driving performance; were a significant predictor of recent crash involvement; and were able to distinguish between crashers and noncrashers; (Richardson & Unroe, 2004; MacGregor, Freeman, & Zhang, 2001; Myers, et al., 2000; Stutts et al.,1998).

PSYCHOMOTOR EVALUATION TOOLS

Movement Speed

Grooved Pegboard Test

The Grooved Pegboard Test (GPT) consists of a board with 25 holes (with randomly positioned slots) and matching pegs. Pegs with a key along one side must be rotated to match the hole before they can be inserted. This test requires more complex visual-motor coordination than most pegboard tests, and has been utilized as part of larger batteries in driving simulator studies (Schanke & Sundet, 2000; Uc et al., 2006). In one study, GPT scores were not found to be significantly different between groups identified as "not suited for driving" and "suited." In another study, logistic regression models calculated odds ratios across groups, and found a significant correlation between GPT scores and unsafe driving simulator actions.

Symbol Digit Modalities Test

The Symbol Digit Modalities Test involves a substitution task, where the examinee has 90 seconds to pair specific numbers with given geometric figures. Responses can be written or oral, and the administration time is 5 minutes. Two separate studies have utilized this measure and compared the results to on-road driving tests

(Schanke & Sundet, 2000; Richardson & Marottoli, 2003). A significant relationship was found in one study, but no significant correlation was found in the other.

RANGE OF MOTION

Foot Tap Test

The foot tap test is the number of seconds it takes a seated subject to tap his or her foot left to right (either over an open notebook or between two circles on the ground) a set number of times. The foot tap test is part of the Gross Impairment Screening Battery (GRIMPS). In a prospective analysis of about 1,000 seniors, no significant correlation was found between three chair stand scores and crashes, citations, or police stops over the 1-year study period. In a prospective cohort study of nearly 2,000 seniors, it was found that the foot tap test scores did predict future crash involvement (Ball et al., 2005; Marottoli et al., 1994).

Neck Flexibility

Neck flexibility refers to the amount a subject can turn his or her neck while keeping the torso facing forward. It can be measured by a graduated scale through observation or by pass/fail for having a subject read the time on a clock placed directly behind him or her. A study of about 100 older drivers found that neck flexibility scores did not significantly correlate with on-road driving performance scores. In a retrospective analysis of about 1,000 seniors, no significant correlation was found between neck flexibility scores and crashes, citations, or police stops over the 1-year study period (Ball et al., 2005; Hunter-Zaworski, 1990; Janke & Eberhard, 1998; Laux & Brelsford, 1990).

Rapid Pace Walk

The rapid pace walk measures the number of seconds it takes a subject to walk 10 feet, turn around, and walk back as quickly as he or she can safely do so. In a prospective analysis of about 1,000 seniors, time on the rapid pace walk (>7 sec vs. ≤ 7 sec) was strongly associated with crashes, citations, and police stops over the 1-year study period (Ball et al., 2005). Studies have found that those who take more than 7 seconds on this test have an increased relative crash risk of 1.25 to 2.0. In a prospective cohort study of nearly 2,000 seniors, it was found that the foot tap test scores did predict future at-fault crash involvement (Marottoli et al., 1994; McCarthy & Mann, 2006).

Arm Reach Test

The arm reach test requires a seated subject to raise each arm, one at a time, above the head. To pass, a subject must be able to raise his or her elbow above the height of the shoulder. This test is a measure of upper limb mobility. In a prospective cohort study of nearly 2,000 seniors, it was found that the arm reach scores did not predict future at-fault crash involvement (Ball et al., 2005). In this study, nearly all subjects passed the test.

REACTION TIME

Simple Reaction Time Test

Simple reaction time is the number of milliseconds it takes a subject to respond to a noncued stimulus. Simple reaction time scores have been compared with driving on road and in a simulator, and with self-reported crashes and violations. No study has found significant associations (Guerrier et al., 1999; Laux & Brelsford, 1990; Lundqvist et al., 2000; Lundberg et al., 1998; Myers et al., 2000; Ranney & Pulling, 1990; Schanke & Sundet, 2000).

Doron Cue Recognition

The Doron Cue Recognition test is administered using a Doron Driving Analyzer, which is a low-fidelity driving simulator. The subject sees car icons on the screen. When one of the icons changes position on the screen, the subject is requested to release the simulator's throttle and press the brake. The score is the number of milliseconds from the icon moving to the release of the throttle, converted to the number of feet the simulated vehicle would have traveled if it were moving at 55 mph. A similar measure can be taken for the steering reaction time, and choice reaction time is measured by having the subject choose appropriately between responses. All three Doron Cue Recognition measures have been found to be significantly correlated with performance on a driving test (Janke & Eberhard, 1998; Janke, 2001).

UPPER/LOWER BODY MUSCLE STRENGTH

Manual Muscle Test

The Manual Muscle Test (MMT) involves the physical testing of muscle strength and flexibility. It is generally used with the shoulders, hands, hip, and knees. Although several studies have compared MMT scores to driving outcomes, the results are mixed (Kantor et al., 2004; Marottoli et al., 1994; McCarthy & Mann, 2006). Work examining only upper body strength found that upper body strength measures did not correlate with on-road driving performance. A prospective cohort study found that upper body strength was not related to adverse outcomes (crashes, citations, being stopped by police) but lower body strength was.

Dynamometer Grip Strength

Grip strength is usually measured with a dynamometer, where the subject squeezes two bars together and strength is measured in lb/in^2. One retrospective analysis found that grip strength was correlated with poorer on-road driving performance. A case control study and a prospective study found no relationship between grip strength and all crashes or at-fault crashes. The evidence suggests that grip strength may not be a good measure of adverse driving performance (Laux & Brelsford, 1990; Sims, McGwin et al., 2000; Sims, Owsley et al., 1998; Kantor et al., 2004).

References

AAA Foundation for Traffic Safety. (1994). *Drivers 55 Plus: Check Your Own Performance.* Washington, DC: AAA Foundation for Traffic Safety.

AAA Foundation for Traffic Safety. (1997). *The Older and Wiser Driver.* Washington, DC: AAA Foundation for Traffic Safety.

AAA Foundation for Traffic Safety. (2006). *How to Help an Older Driver: A Guide for Planning Safe Transportation.* Washington, DC: AAA Foundation for Traffic Safety.

Abdel-Aty, M.A., Chen, C.L., & Schott, J.R. (1998). An assessment of the effect of driver age on traffic accident involvement using log-linear models. *Accident Analysis & Prevention,* 30, 851–861.

Abdel-Aty, M.A. & Radwan, A.E. (2000). Modeling traffic accident occurrence and involvement. *Accident Analysis & Prevention,* 32, 633–642.

Abel, L.A., Troost, B.T., & Dell'Osso, L.F. (1983). The effects of age on normal saccadic characteristics and their variability. *Vision Research,* 23, 33–37.

Adler, G., Bauer, M., Rottunda, S., & Kuskowitz, M. (2005). Driving habits and patterns in older men with glaucoma. *Social Work in Health Care,* 40, 75–87.

Adler, G. & Kuskowski, M. (2003). Driving cessation in older men with dementia. *Alzheimer Disease & Associated Disorders,* 17, 68–71.

Adler, G. & Rottunda, S. (2006). Older adults' perspectives on driving cessation. *Journal of Aging Studies,* 20, 227–235.

Adler, G., Rottunda, S.J., & Dusken, M.W. (1996). The driver with dementia. *American Journal of Geriatric Psychiatry,* 4, 110–120.

Alan M. Voorhees Transportation Center. (2005). *Safe Mobility at Any Age: Policy Forum Series* (Final Report). New Brunswick, NJ: Author and New Jersey Foundation for Aging.

Alexander, G.J. & Lunenfeld, H. (1986). *Driver Expectancy in Highway Design and Traffic Operations.* (Report No. FHWA-TO-86-1). Washington, DC: Federal Highway Administration.

Al-Madani, H. & Al-Janahi, A. (2002). Assessment of drivers' comprehension of traffic signs based on their traffic, personal and social characteristics. *Transportation Research Part F,* 5, 63–76.

Alvarez, F.J. & Fierro, I. (2008). Older drivers, medical conditions, medical impairment and crash risk. *Accident Analysis & Prevention,* 40, 55–60.

American Association of Motor Vehicle Administrators. (2006). *Challenging Myths and Opening Minds Forum: Aging and the Medically At-Risk Driver.* URL: http://www.aamva.org/Events/Materials/2006ChallengingMythsRecap.htm

American Association of Retired Persons. (1992). *Creating Mobility Choices: Older Driver Skill Assessment and Resource Guide.* Washington, DC: AARP.

American Automobile Association. (2005). *AAA Mid-Atlantic Digest of Motor Laws.* Available at http://www.aaamidatlantic.com/automotive/MotorLaws/. Accessed April 12, 2005.

American Diabetes Association. (2008). *All About Diabetes: Total Prevalence of Diabetes & Pre-Diabetes.* URL: http://www.diabetes.org.

American Heritage Dictionary. (2000). *The American Heritage® Dictionary of the English Language, Fourth Edition.* Houghton Mifflin Company.

American Occupational Therapy Association. (2002a). Occupational therapy practice framework: Domain and process. *American Journal of Occupational Therapy,* 56, 609–639.

American Occupational Therapy Association. (2002b). *Keeping Older Drivers Safe.* Washington, DC: American Occupational Therapy Association.

Anders, T.R., Fozard, J.L., & Lillyquist, T.D. (1972). Effects of age upon retrieval from short-term memory. *Developmental Psychology,* 6, 214–217.

Andersen, G.J. & Atchley, P. (1995). Age-related differences in the detection of three-dimensional surfaces from optic flow. *Psychology and Aging*, 10, 650–658.

Anderson, J.R. (1985). *Cognitive Psychology and Its Implications*. New York, NY: Freeman and Company.

Anshel, M.H. (1978). Effect of aging on acquisition and short-term retention of a motor skill. *Perceptual Motor Skills*, 47, 993–994.

Anstey, K.J., Windsor, T.D., Luszcz, M.A., & Andrews, G.R. (2006). Predicting driving cessation over 5 years in older adults: Psychological well-being and cognitive competence are stronger predictors than physical health. *Journal of the American Geriatrics Society*, 54, 121–126.

Anstey, K.J., Wood, J., Lord, S., & Walker, J.G. (2005). Cognitive, sensory and physical factors enabling driving safety in older adults. *Clinical Psychology Review*, 25, 45–65.

Aparasu, R.R. & Mort, J.R. (2004). Prevalence, correlates, and associated outcomes of potentially inappropriate psychotropic use in the community-dwelling elderly. *The American Journal of Geriatric Pharmacotherapy*, 2, 102–111.

Arditi, A. (1986). Binocular vision. In K.R. Boff, L. Kaufman, & J.P. Thomas (Eds.), *Handbook of Perception and Human Performance: Vol. 1. Sensory Processes and Perception*. New York, NY: John Wiley and Sons.

Arenberg, D. (1976). The effects of input condition on free recall in young and old adults. *Journal of Gerontology*, 31, 551–555.

Association of Driver Rehabilitation Specialists. (2002). *Model Practices for Driver Rehabilitation for Individuals With Disabilities*. Hickory, NC: Association of Driver Rehabilitation Specialists, ADED.

Association of Driver Rehabilitation Specialists. (2004). *Best Practices for the Delivery of Driver Rehabilitation Services, 2004*. Hickory, NC: Association of Driver Rehabilitation Specialists, ADED.

Asthma and Allergy Foundation of America. (2008). *Asthma Overview*. URL: http://www.aafa.org/.

Attebo, K., Mitchell, P., & Smith, W. (1996). Visual acuity and the causes of visual loss in Australia: The Blue Mountain Eye Study. *Ophthalmology*, 103, 357–364.

Aubrey, J.B., Li, K.Z., & Dobbs, A.R. (1994). Age differences in the interpretation of misaligned "you-are-here" maps. *Journal of Gerontology: Psychological Sciences*, 49, 29–31.

Austroads. (2006). *Assessing Fitness to Drive for Commercial and Private Vehicle Drivers*. Sydney, Australia: Austroads Inc.

Avolio, B.J., Kroeck, K.G., & Panek, P.E. (1985). Individual differences in information-processing ability as a predictor of motor vehicle accidents. *Human Factors*, 27, 577–587.

Axtell, K.A. & Akhtar, M. (1990). Incidence of syncope prior to implantable cardioverter defibrillator discharge. *Circulation*, 82, 211.

Azad, N., Byszewski, A., Amos, S., & Molnar, F.J. (2002). A survey of the impact of driving cessation on older drivers. *Geriatrics Today*, 5, 170–174.

Baddeley, A.D. (1984). The fractionation of human memory. *Psychological Medicine*, 14, 259–264.

Bacon, F. (1597). *Meditationes Sacrae*. London, England: John Windet for Humfrey Hooper.

Bahrick, H.P. (1984). Memory for older people. In J.E. Harris & P.E. Morris (Eds.), *Everyday Memory, Actions and Absentmindedness*. New York, NY: Academic Press.

Bahrick, H.P., Bahrick, P.O., & Wittlinger, R.P. (1975). Fifty years of memory for names and faces: A cross-sectional approach. *Journal Experimental Psychology*, 104, 54–75.

Bailey, I.L., & Sheedy, J.E. (1988). Vision screening for driver licensure. In *Transportation in an Aging Society: Improving Mobility and Safety for Older Persons, Special Report 218: Volume 2*. Washington, DC: Transportation Research Board.

Bailey, L. (2004). *Aging Americans: Stranded Without Options*. Washington, DC: Surface Transportation Policy Project.

Baker, T.K., Falb, T., Voas, R., & Lacey, J. (2003). Older women drivers: Fatal crashes in good conditions. *Journal of Safety Research*, 34, 399–405.

Baldock, M.R.J., Mathias, J.L., McLean, A.J., & Berndt, A. (2006). Self-regulation of driving and its relationship to driving ability among older adults. *Accident Analysis & Prevention*, 38, 1038–1045.

Ball, K. (1997). Attentional problems and older drivers. *Alzheimer Disease & Associated Disorders*, 11, 42–47.

Ball, K. & Owsley, C. (1991). Identifying correlates of accident involvement for the older driver. *Human Factors*, 33, 583–595.

Ball, K. & Rebok, G. (1994). Evaluating the driving ability of older adults. *Journal of Applied Gerontology*, 13, 20–38.

Ball, K. & Sekuler, R. (1986). Improving visual perception in older observers. *Journal of Gerontology*, 41, 176–182.

Ball, K.K., Beard, B.L., Roenker, D.L., Miller, R.L., & Griggs, D.S. (1988). Age and visual search: Expanding the useful field of view. *Journal of the Optical Society of America*, 5, 2210–2219.

Ball, K.K., Berch, D.B., Helmer, K.F., Jobe, J.B., Leveck, M.D., Marsiske, M., Morris, J.N., Rebok, G.W., Smith, D.M., Tennstedt, S.L., Unverzagt, F.W., & Willis, S.L. (2007). Effects of cognitive training interventions with older adults: A randomized control trial. *Journal of the American Medical Association*, 288, 2271–2281.

Ball, K.K., Owsley, C., Sloane, M.E., Roeneker, D.L., & Bruni, J.R. (1993). Visual attention problems as a predictor of vehicle crashes in older drivers. *Investigative Ophthalmology & Visual Science*, 34, 3110–3123.

Ball, K.K., Owsley, C., Stalvey, B., Roenker, D.L., Sloane, M.E., & Graves, M. (1998). Driving avoidance and functional impairment in older drivers. *Accident Analysis & Prevention*, 30, 313–322.

Ball, K.K., Roenker, D.L., Wadley, V.G., Edwards, J.D., Roth, D.L., McGwin, G., Jr., Raleigh, R., Joyce, J.J., Cissell, G.M., & Dube, T. (2005). Can high-risk older drivers be identified through performance-based measures in a Department of Motor Vehicles setting? *The American Geriatrics Society*, 54, 77–84.

Bansch, D., Brunn, J., Castrucci, M., Weber, M., Gietzen, F., Borggrefe, M., Breithardt, G., & Block, M. (1998). Syncope in patients with an implantable cardioverter defibrillator: Incidence, prediction, and implications for driving restrictions. *Journal of the American College of Cardiology*, 31, 608–615.

Barbé, F., Pericás, J., Muñoz, A., Findley, L., Antó, J.M., & Agustí, A.G.N. (1998). Automobile accidents in patients with sleep apnea syndrome. *American Journal of Respiratory Critical Care Medicine*, 158, 18–22.

Barrett, G.V., Mihal, W.L., Panek, P.E., Sterns, H.L., & Alexander, R.A. (1977). Information-processing skills predictive of accident involvement for younger and older commercial drivers. *Industrial Gerontology*, 4, 173–182.

Bauer, M.J., Rottunda, S., & Adler, G. (2003). Older women and driving cessation. *Qualitative Social Work*, 2, 309–325.

Bédard, M., Isherwood, I., Moore, E., Gibbons, C., & Lindstrom, W. (2004). Evaluation of a re-training program for older drivers. *Canadian Journal of Public Health*, 95, 295–298.

Bédard, M., Montplaisir, J., Richer, F., & Malo, J. (1991). Nocturnal hypoxemia as a determinant of vigilance impairment in sleep apnea syndrome. *Chest*, 100, 367–371.

Bédard, M., Riendeau, J., & Weaver, B. (2007). The relationship between driving performance and scores on the Roadwise Review. *The Gerontologist*, 47, 73.

Beers, M.H. (1997). Explicit criteria for determining potentially inappropriate medication use by the elderly: An update. *Archives of Internal Medicine*, 157, 1531–1537.

Bell, M.D., Wolf, E., & Bernholz, C.D. (1972). Depth perception as a function of age. *Aging & Human Development*, 3, 77–81.

Bellis, M. (2008a). *History of Aspirin*. URL: http://inventors.about.com/library/inventors/blasprin.htm.

Bellis, M. (2008b). *Automobile History*. URL: http://inventors.about.com/library/inventors/blcar.htm.

Ben-Bassat, T. & Shinar, D. (2006). Ergonomic guidelines for traffic sign design increase sign comprehension. *Human Factors*, 48, 182–195.

Benekohal, R.F., Michaels, R.M., Shim, E., & Resende, P.T.V. (1994). Effects of aging on older drivers' travel characteristics. *Transportation Research Record*, 1438, 91–98.

Benekohal, R.F., Resende, P., Shim, E., Michaels, R.M., & Weeks, B. (1992). *Highway Operations Problems of Elderly Drivers in Illinois*. (Report No. FHWA-IL-92-023). Springfield, IL: Illinois Department of Transportation.

Berg, C.A., Hertzog, C., & Hunt, E. (1982). Age differences in the speed of mental rotation. *Developmental Psychology*, 18, 95–107.

Berg, H.Y. (2006). Reducing crashes and injuries among young drivers: What kind of prevention should we be focusing on? *Injury Prevention*, 12(Suppl), i15–i18.

Berghaus, G. & Grass, H. (1997). Concentration-effect relationship with benzodiazepine therapy. In C. Mercier-Guyon (Ed.), *Proceedings—14th International Conference on Alcohol, Drugs, and Traffic Safety*. Annecy, France: Centre d'Etudes et de Recherchers en Médeccine du Traffic.

Bernstein, D.A., Roy, E.J., Srull, T.K., & Wickens, C.D. (1991). *Psychology, 2nd Edition*. Boston, MA: Houghton Mifflin Company.

Berryman, M.E. (2004). Automatic crash notification and 9-1-1: A success story. In *Proceedings of the 2004 ESRI Users Conference*. Redlands, CA: ESRI International.

Berube, E. (1995). *Evaluation of Mature Driver Improvement Program Home-Study Courses*. (Report No. RSS-95-157). Sacramento, CA: California Department of Motor Vehicles.

Betts, T., Markman, D., Debenham, S., Mortiboy, D., & McKevitt, T. (1984). Effects of two antihistamine drugs on actual driving performance. *British Medical Journal*, 288, 281–282.

Beverly Foundation. (2004a). *Innovations for Seniors: Public and Community Transit Services Respond to Special Needs*. Pasadena, CA: Beverly Foundation and the Community Transportation Association of America.

Beverly Foundation. (2004b). *Supplemental Transportation Programs for Seniors: A Report on STPs in America*. Pasadena, CA: Beverly Foundation.

Bianchi, A. & Summala, H. (2004). The "genetics" of driving behavior: Parents' driving style predicts their children's driving style. *Accident Analysis & Prevention*, 36, 655–659.

Bingham, C.R., Eby, D.W., Hockanson, H.W., & Greenspan, A.I. (2006). Factors influencing the use of booster seats: A statewide survey of parents. *Accident Analysis & Prevention*, 38, 1028–1037.

Birren, J.E. & Shock, N.W. (1950). Age changes in rate and level of dark adaptation. *Journal of Applied Psychology*, 26, 407–411.

Blazer, D.G., Hybels, C.F., Simonsick, E.M., & Hanlon, J.T. (2000a). Marked differences in antidepressant use by race in an elderly community sample: 1986–1996. *American Journal of Psychiatry*, 157, 1089–1094.

Blazer, D.G., Hybels, C.F., Simonsick, E.M., & Hanlon, J.T. (2000b). Sedative, hypnotic, and anti-anxiety medication use in an aging cohort over ten years: A racial comparison. *Journal of the American Geriatric Society*, 48, 1073–1079.

Bleakley, J.F. & Akiyama, T. (2003). Driving and arrhythmias: Implications of new data. *Cardiac Electrophysiology Review*, 7, 77–79.

Blendon, R.J., DesRoches, C.M., Benson, J.M., Brodie, M., & Altman, D.E. (2001). Americans' views on the use and regulation of dietary supplements. *Archives of Internal Medicine,* 16, 805–810.

Boettner, E.A. & Wolter, J.R. (1962). Transmission of the ocular media. *Investigative Ophthalmology & Visual Science,* 1, 776–783.

Bogner, H.R., Straton, J.B., Gallo, J.J., Rebok, G.W., & Keyl, P.M. (2004). The role of physicians in assessing older drivers: Barriers, opportunities, and strategies. *Journal of the American Board of Family Practice,* 17, 38–43.

Bonema, J.D. & Maddens, M.E. (1992). Syncope in elderly patients: Why their risk is higher. *Postgraduate Medicine,* 91, 129–144.

Bouman, J. & Pellerito, J.M. (2006). Preparing for the on-road evaluation. In J.M. Pellerito (Ed.), *Driver Rehabilitation and Community Mobility.* St. Louis, MO: Elsevier Mosby.

Bowers, A., Peli, E., Elgin, J., McGwin, G. Jr., & Owsley, C. (2005). On-road driving with moderate visual field loss. *Optometry and Vision Science,* 82, 657–667.

Braitman, K.A., Kirley, B.B., Ferguson, S., & Chaudhary, N.K. (2007). Factors leading to older drivers' intersection crashes. *Traffic Injury Prevention,* 8, 267–274.

Brancato, R. (1969). Il tempo di recupero in seguito ad abbagliamento in funzione dell'età. *Atti della "Fondazione Georgio Ronchi,"* 24, 585–588.

Braunstein, M.L. (1986). Perception of rotation in depth: The psychophysical evidence. In N.I. Badler & J.K. Tsotsos (Eds.), *Motion: Representation and Perception.* New York, NY: North Holland.

Brody, B.L., Gamst, A.C., Williams, R.A., Smith, A.R., Lau, P.W., Dolnak, D., Rapaport, M.H., Kaplan, R.M., & Brown, S.I. (2001). Depression, visual acuity, comorbidity, and disability associated with age-related macular degeneration. *Ophthalmology,* 108, 1893–1901.

Brown, B. & Bowman, K.J. (1987). Sensitivity to changes in size and velocity in young and elderly observers. *Perception,* 16, 41–47.

Brown, J.A. (1958). Some tests of the decay theory of immediate memory. *Quarterly Journal of Experimental Psychology,* 10, 12–21.

Brown, L.B., Ott, B.R., Papandonatos, G.D., Sui, Y., Ready, R.E., & Morris, J.C. (2005). Prediction of on-road driving performance in patients with early Alzheimer's disease. *Journal of the American Gerontological Society,* 53, 94–98.

Bruff, J.T., & Evans, J. (1999). *Elderly Mobility and Safety—The Michigan Approach Final Plan of Action.* Retrieved March 24, 2003, from http://www.semcog.org/Products/pdfs/eldmob_final.pdf

Bruno, J.J. & Ellis, J.J. (2005). Herbal use among US elderly: 2002 National Health Interview Survey. *The Annals of Pharmacotherapy,* 39, 643–648.

Bulstrode, S.J. (1987). Car mirrors for drivers with restricted neck mobility. *International Disabilities Studies,* 9, 180–181.

Burg, A. (1966). Visual acuity as measured by dynamic and static tests. *Journal of Applied Psychology,* 50, 460–466.

Burg, A. (1968). Lateral visual field as related to age and sex. *Journal of Applied Psychology,* 52, 10–15.

Burg, A. & Hurlbert, S. (1961). Dynamic visual acuity as related to age, sex, and static acuity. *Journal of Applied Psychology,* 45, 111–116.

Burkhardt, J.E. (2003). Better transportation services for older persons. *Transportation Research Record,* 1843, 105–112.

Burkhardt, J.E. & McGavock, A.T. (1999). Tomorrow's older drivers: Who? How many? What impacts? *Transportation Research Record,* 1693, 62–70.

Burkhardt, J.E., McGavock, A.T., & Nelson, C.A. (2002). *Improving Public Transit Options for Older Persons: Volume 1. Handbook.* (Report No. 82). Washington, DC: Transportation Research Board.

Burkhardt, J.E., McGavock, A.T., Nelson, C.A., & Mitchell, C.G.B. (2002). *Improving Public Transit Options for Older Persons: Volume 2. Final Report* (Report No. 82). Washington, DC: Transportation Research Board.

Byszewski, A.M., Graham, I.D., Amos, S., Man-Son-Hing, M., Dalziel, W.B., Marshall, S., Hunt, L., Bush, C., & Guzman, D. (2003). A continuing medical education initiative for Canadian primary care physicians: The Driving and Dementia Toolkit: A pre-and post-evaluation of knowledge, confidence gained, and satisfaction. *Journal of American Geriatric Society*, 51, 1484–1489.

Cadigan, D.A., Magaziner, J., & Fedder, D.O. (1989). Polymedicine use among community resident older women: How much of a problem? *American Journal of Public Health*, 79, 1537–1540.

Caird, J.K. (2004). In-vehicle intelligent transportation systems: Safety and mobility of older drivers. In *Transportation in an Aging Society: A Decade of Experience*. Washington, DC: Transportation Research Board.

Caird, J.K., Chugh, J.S., Wilcox, S., & Dewar, R.E. (1998). *A Design Guideline and Evaluation Framework to Determine the Relative Safety of In-Vehicle Intelligent Transportation Systems for Older Drivers.* (Report No. TP 13349E). Montreal, Quebec: Transportation Development Centre.

Campbell, M.K., Bush, T.L., & Hale, W.E. (1993). Medical conditions associated with driving cessation in community-dwelling, ambulatory elders. *Journal of Gerontology: Social Sciences*, 48, S230–S234.

Canadian Cardiovascular Society. (1996). Assessment of the cardiac patient for fitness to drive: 1996 update. *Canadian Journal of Cardiology*, 12, 1164–1170.

Canadian Medical Association. (2006). *Determining Medical Fitness to Operate Motor Vehicles, 7th Edition.* Ottawa, Canada: Canadian Medical Association.

Carberry, T.P., Wood, J.M., & Watson, B.C. (2004). Eye diseases and driving performance: Correlates between insight and capability in drivers with glaucoma. In *2004 Road Safety Research, Policing, and Education Conference.* Perth, Western Australia.

Carp, F.M. (1988). Significance of mobility for the well-being of the elderly. In *Transportation in an Aging Society: Improving Mobility and Safety of Older Persons, Volume 2.* Washington, DC: Transportation Research Board.

Carr, D.B. (2000). The older adult driver. *American Family Physician*, 61, 141–146.

Carr, D.B. (2004). Commentary: The role of the emergency physician in older driver safety. *Annals of Emergency Medicine*, 43, 747–748.

Carr, D.B. (2008). Current knowledge on medical fitness to drive: The role of the clinician. In D.W. Eby & L.J. Molnar (Eds.), *Proceedings of the North American License Policies Workshop.* Washington, DC: AAA Foundation for Traffic Safety.

Carr, D.B., Jackson, T., & Alguire, P. (1990). Characteristics of an elderly driving population referred to a geriatric assessment center. *Journal of the American Geriatric Society*, 38, 1145–1150.

Carr, D.B., LaBarge, E., Dunningan, K., & Storandt, M. (1998). Differentiating drivers with dementia of the Alzheimer type from healthy older persons with a traffic sign naming test. *The Journals of Gerontology*, 53A, M135–M139.

Carr, R.W. (2000). Editorial in *Ann Arbor News*. January 25, 2000.

Centers for Disease Control and Prevention. (1994). Prevalence of self-reported epilepsy: United States, 1986–1990. *Morbidity and Mortality Weekly Report*, 43, 810–811, 817–818.

Centers for Disease Control and Prevention. (1997). *National Diabetes Fact Sheet*. Atlanta, GA: National Center for Chronic Disease Prevention and Health Promotion.

Centers for Disease Control and Prevention. (2003). Life expectancy at birth, at 65 years of age, and at 75 years of age, according to race and sex: United States, selected years 1900–2001. *National Vital Statistics Reports*, 52, 133.

Centers for Disease Control and Prevention. (2004). *Preventing Heart Disease and Stroke: Addressing the Nation's Leading Killers, 2004*. Atlanta, GA: Centers for Disease Control and Prevention.

Centers for Disease Control and Prevention. (2006). Prevalence of doctor-diagnosed arthritis and arthritis-attributable activity limitation—United States, 2003–2005. *Morbidity and Mortality Weekly Report*, 55, 1089–1092.

Cerella, J. (1985). Information processing rates in the elderly. *Psychology Bulletin*, 98, 67–83.

Cerella, J., Poon, L.W., & Fozard, J.L. (1981). Mental rotation and age reconsidered. *Journal of Gerontology*, 38, 447–454.

Chamberlain, W. (1970). Restriction of upward gaze with advancing age. *Transactions of the American Ophthalmological Society*, 68, 234–244.

Chamberlain, W. (1971). Restriction of upward gaze with advancing age. *American Journal of Ophthalmology*, 71, 341–346.

Champion, H.R., Augenstein, J.S., Blatt, A.J., Cushing, B., Digges, K.H., Hunt, R.C., Lombardo, L.V., & Siegel, J.H. (2003). Reducing highway deaths and disabilities with automatic wireless transmission of serious injury probability ratings from vehicles in crashes to EMS. In *Proceedings of the 18th International Technical Conference on the Enhanced Safety of Vehicles*. Washington, DC: U.S. Department of Transportation.

Chandraratna, S. & Stamatiadis, N. (2003). Problem driving maneuvers of elderly drivers. *Transportation Research Record*, 1843, 89–95.

Chantelau, E. (1991). Epilepsy or diabetes and auto accidents: Letter to the editor. *New England Journal of Medicine*, 324, 1510.

Charlton, J., Koppel, S., O'Hare, M., Andrea, D., Smith, G., Khodr, B., Langford, J., Odell, M., & Fildes, B. (2004). *Influence of Chronic Illness on Crash Involvement of Motor Vehicle Drivers*. (Report No. 213). Victoria, Australia: Monash University Accident Research Centre.

Charlton, J.L., Oxley, J., Fildes, B., & Les, M. (2001). *Self-Regulatory Behaviour of Older Drivers*. Paper presented at the Road Safety Research, Policing and Education Conference, Melbourne, Victoria, Australia.

Charlton, J.L., Oxley, J., Fildes, B., Oxley, P., Newstead, S., Koppel, S., & O'Hare, M. (2006). Characteristics of older drivers who adopt self-regulatory driving behaviors. *Transportation Research Part F*, 9, 363–373.

Chipman, M.L., MacGregor, C.G., Smiley, A.M., & Lee-Gosselin, M. (1993). The role of exposure in comparisons of crash risk among different drivers and driving environments. *Accident Analysis & Prevention*, 25, 207–211.

Chiu, S.A., Mourant, R., & Bond, R. (1997). *A Study of Driver Behavior in Highway Construction Zones With Respect to Age and Illumination Conditions*. Paper presented at the 76th Annual Meeting of the Transportation Research Board. Washington, DC.

Chrischilles, E.A., Lemke, J.H., Wallace, R.B., & Drube, G.A. (1990). Prevalence and characteristics of multiple drug use in an elderly study group. *Journal of the American Geriatrics Society*, 38, 979–984.

Clark, D.D., Forsyth, R., & Wright, R. (1999). Junction road accidents during cross-flow turns: A sequence analysis of police case files. *Accident Analysis & Prevention*, 31, 31–43.

Clark, D.E. & Cushing, B.M. (2002). Predicted effect of automatic crash notification on traffic mortality. *Accident Analysis & Prevention*, 34, 507–513.

Clarke, W.L., Cox, D.J., Gonder-Frederick, L.A., & Kovatchev, B. (1999). Hypoglycemia and the decision to drive a motor vehicle by persons with diabetes. *Journal of the American Medical Association*, 282, 750–754.

Classen, S. & Awadzi, K. (2008). Model state programs on licensing older drivers. In D.W. Eby & L.J. Molnar (Eds.), *Proceedings of the North American License Policies Workshop*. Washington, DC: AAA Foundation for Traffic Safety.

Classen, S., Shechtman, O., Stephens, B., Bendixen, R., Sandhu, M., McCarthy, D., & Mann, W. (2006). The impact of roadway intersection design on driving performance of young and senior adults: Preliminary results. *Topics in Geriatric Rehabilitation*, 22, 18–26.

Classen, S., Shechtman, O., Stephens, B., Davis, E., Justiss, M., Bendixen, R., Belchoir, P., Sandhu, M., Posse, C., McCarthy, D., & Mann, W. (2007). Impact of roadway intersection design on driving performance of young and senior adults. *Traffic Injury Prevention*, 8, 69–77.

Clayton, A.B., Harvey, P.G., & Betts, T.A. (1977). The effects of two antidepressants, imiparmine and viloxazine, upon driving performance. *Psychopharmacology*, 55, 9–12.

Coeckelbergh, T.R.M., Brouwer, W.H., Cornelissen, F.W., van Woffelaar, P., & Kooilman, A.C. (2002). The effects of visual field defects on driving performance: A driving simulator study. *Archives of Ophthalmology*, 120, 1509–1516.

Coeckelbergh, T.R.M., Cornelissen, F.W., Brouwer, W.H., & Kooijman, A.C. (2002). The effect of visual field defects on eye movements and practical fitness to drive. *Vision Research*, 42, 669–677.

Collia, D. V. Sharp, J., & Giesbrecht, L. (2003). The 2001 National Household Travel Survey: A look into the travel patterns of older Americans. *Journal of Safety Research*, 34(4), 461–470.

Cook, L.J., Knight, S., Olson, L.M., Nechodom, P.J., & Dean, J.M. (2000). Motor vehicle crash characteristics and medical outcomes among older drivers in Utah, 1992–1995. *Annals of Emergency Medicine*, 35, 585–591.

Cooper, P. (1990). Differences in accident characteristics among elderly drivers and between elderly and middle-aged drivers. *Accident Analysis & Prevention*, 5, 499–508.

COPD International. (2004). *COPD Statistics*. URL: http://www.copd-international.com/library/statistics.htm.

Coren, S. & Girgus, J.S. (1972). Density of human lens pigment: In vivo measure over an extended age range. *Vision Research*, 12, 343–346.

Cornelius, S.W. & Caspi, A. (1987). Everyday problem solving in adulthood and old age. *Psychology & Aging*, 2, 144–153.

Cornoni-Huntley, J., Brock, D.B., Ostfeld, A.M., Taylor, J.O., & Wallace, R.B. (1986). *Established Populations for Epidemiological Study of the Elderly: Resource Data Book* (NIH Publication No. 86-2443). Bethesda, MD: National Institute on Aging.

Corso, J.F. (1981). *Aging Sensory Systems and Perception*. New York, NY: Praeger.

Cotté, N., Meyer, J., & Coughlin, J.F. (2001). Older and younger drivers' reliance on collision warning systems. In *Proceedings of the Human Factors and Ergonomics Society 45th Annual Meeting*. Santa Monica, CA: Human Factors and Ergonomics Society.

Coughlin, J.F. (2001). *Beyond Health and Retirement: Placing Transportation on the Aging Policy Agenda*. Cambridge, MA: Massachusetts Institute of Technology, Center for Transportation Studies, Age Lab.

Coughlin, J.F. (2005). Not your father's auto industry? Aging, the automobile, and the drive for product innovation. *Generations*, 28(4), 38–44.

Coughlin, J.F. & Lacombe, A. (1997). Ten myths about transportation for the elderly. *Transportation Quarterly*, 51, 91–100.

Coughlin, J.F., Mohyde, M., D'Ambrosio, L.A., & Gilbert, J. (2004). *Who Drives Older Driver Decisions?* Cambridge, MA: MIT AgeLab.

Cox, D.J., Clarke, W.L., Gonder-Frederick, L.A., & Koatchev, B.P. (2001). Driving mishaps and hypoglycemia: Risk and prevention. *International Journal of Clinical Practice Supplement*, 123, 38–42.

Cox, D.J., Gonder-Frederick, L.A., & Clarke, W.L. (1993). Driving decrements in type I diabetes during moderate hypoglycemia. *Diabetes*, 42, 239–243.

Cox, D.J., Gonder-Frederick, L.A., Kovatchev, B.P., Julian, D.M., & Clarke, W.L. (2000). Progressive hypoglycemia's impact on driving simulation performance. *Diabetes Care*, 23, 163–170.

Cox, D.J., Quillian, W.C., Thorndike, F.P., et al. (1998). Evaluating driving performance of outpatients with Alzheimer's Disease. *Journal of the American Board of Family Medicine*, 11, 264–271.

Craik, F.I.M. (1968). Two components in free recall. *Journal of Verbal Learning & Verbal Behavior*, 7, 996–1004.

Crassini, B., Brown, B., & Bowman, K. (1988). Age related changes in contrast sensitivity in the central and peripheral retina. *Perception*, 17, 315–332.

Crook, T.H., West, R.L., & Larrabee, G.J. (1993). The driving-reaction time test: Assessing age declines in dual-task performance. *Developmental Neuropsychology*, 9, 31–39.

Cushman, L.A. (1992). *The Impact of Cognitive Decline and Dementia on Driving in Older Adults*. Washington, DC: AAA Foundation for Traffic Safety.

Cushman, L.A. & Cogliandro, F. (1997). Cognitive capacity: Its relationship to safety in older drivers. *American Driver and Traffic Safety Chronicle*, 45, 8–11.

Cutler, S.J. & Grams, A.E. (1988). Correlates of self-reported everyday memory problems. *Journal of Gerontology: Social Sciences*, 43, 82–90.

D'Ambrosio, L.A., Coughlin, J.F., Mohyde, M., & Reimer, B. (2007). Family matter: Older drivers and the driving decision. *Transportation Research Board*, 2009, 23–29.

Daley, K.F. (1981). Roundabouts: A review of accident patterns. *First National Local Government Engineering Conference*. Adelaide, Australia.

Davey, J.A. (2007). Older people and transport: Coping without a car. *Ageing & Society*, 27, 49–65.

DeCarlo, D.K., Scilley, K., Wells, J., & Owsley, C. (2003). Driving habits and health related quality of life in patients with age-related maculopathy. *Optometry and Vision Science*, 80, 207–213.

Decina, L.E. & Staplin, L. (1993). Retrospective evaluation of alternative vision screening criteria for older and younger drivers. *Accident Analysis & Prevention*, 25, 267–275.

Decter, B.M., Goldner, B., & Cohen, T.J. (1994). Vasovagal syncope as a cause of motor vehicle accidents. *American Heart Journal*, 127, 1619–1621.

DeGroot, M., Anderson, R., Freedland, K.E., Clouse, R.E., & Lustman, P.J. (2001). Association of depression and diabetes complications: A meta-analysis. *Psychosomatic Medicine*, 63, 619–630.

Dejeammes, M. & Ramet, M. (1996). Aging process and safety enhancement of car occupants. *Proceedings of the 15th International Technical Conference on the Enhanced Safety of Vehicles*. Melbourne, Australia.

Delahunt, P.B., Hardy, J.L., Brenner, D.F., Chan, S.C., Dewey, J.A., Mahncke, H.W., Wade, T.W., & Merzenich, M.M. (2008). *InSight: Scientific Principles of a Brain-Plasticity-Based Visual Training Program*. San Francisco, CA: Posit Science Corporation.

Dellinger, A.M., Sehgal, M., Sleet, D.A., & Barrett-Connor, E. (2001). Driving cessation: What older former drivers tell us. *Journal of the American Geriatrics Society*, 49, 431–435.

Demick, J. & Harkins, D. (1999). Cognitive style and driving skills in adulthood: Implications for licensing of older drivers. *International Association of Traffic Safety and Science Research*, 23, 42–57.

Denney, N.W. & Denney, D.R. (1974). Modeling effects on the questioning strategies of the elderly. *Developmental Psychology*, 10, 458.

Denney, N.W. & Palmer, M. (1981). Adult age differences on traditional and practical problem-solving measures. *Journal of Gerontology*, 36, 323–328.

Denney, N.W. & Pearce, K.A. (1989). A developmental study of practical problem solving in adults. *Psychology & Aging*, 4, 438–442.

Denney, N.W., Pearce, K.A., & Palmer, M. (1982). A developmental study of adults' performance on traditional and practical problem-solving tasks. *Experimental Aging Research,*
8, 115–118.

Department of Transport. (2001). *Older Drivers: A Literature Review.* London, UK:
Department of Transport.

DeRaedt, R. & Ponjaert-Kristoffersen, I. (2000). The relationship between cognitive/neuropsy-
chological factors and car driving performance in older adults. *Journal of the American
Geriatrics Society,* 48, 1664–1668.

Derefeldt, G., Lennerstrand, G., & Lundh, B. (1979). Age variation in normal human contrast
sensitivity. *Acta Ophthalmology,* 57, 679–690.

Devos, H., Vandenberghe, W., Nieuwboer, A., Tant, M., Baten, G., & De Weerdt, W. (2007).
Predictors of fitness to driving in people with Parkinson's disease. *Neurology,* 69,
1434–1441.

Dewer, R. & Hanscom, F. (2007). Highway work zones. In R. Dewer & P. Olson (Eds.), *Human
Factors in Traffic Safety.* Tucson, AZ: Lawyers and Judges Publishing Company.

Dewer, R. & Olson, P. (2007). Traffic control devices. In R. Dewer & P. Olson (Eds.), *Human
Factors in Traffic Safety.* Tucson, AZ: Lawyers and Judges Publishing Company.

Dewer, R. (2007). Roadway design. In R. Dewer & P. Olson (Eds.), *Human Factors in Traffic
Safety.* Tucson, AZ: Lawyers and Judges Publishing Company.

Dewer, R.E., Kline, D.W., & Swanson, H.A. (1994). Age differences in comprehension of traffic sign symbols. *Transportation Research Record,* 1456, 1–10.

Dewer, R.E., Kline, D.W., Schieber, F., & Swanson, H.A. (1994). *Symbol Signing Design
for Older Drivers: Final Report.* (Report No. FHWA-RD-94-069). Washington, DC:
Federal Highway Administration.

Di Stefano, M. & Macdonald, W. (2003). Driving instructor interventions during on-road tests
of functionally impaired drivers: Implications for test criteria. In *Proceedings of the
2003 Road Safety Research, Policing and Education Conference.* Sydney, Australia.

Di Stefano, M. & Macdonald, W. (2006). On-the-road evaluation of driving performance.
In J.M. Pellerito (Ed.), *Driver Rehabilitation and Community Mobility: Principles and
Practice.* St. Louis, MO: Elsevier Mosby.

Dickerson, A.E., Molnar, L.J., Eby, D.W., Adler, G., Bédard, M., Berg-Weger, M., Classen,
S., Foley, D., Horowitz, A., Kerschner, H., Page, O., Silverstein, N.M., Staplin, L., &
Trujillo, L. (2007). Transportation and aging: A research agenda for advancing safe
mobility. *The Gerontologist,* 47, 578–590.

Diller, E., Cook, L., Leonard, D., Dean, J.M., Reading, J., & Vernon, D. (1998). *Evaluating
Drivers Licensed with Medical Conditions in Utah, 1992–1996.* (Report No. DOT HS
809 023). Washington, DC: U.S. Department of Transportation.

Dingus, T., Hulse, M.C., Mollnehauer, M.A., Fleischman, R.N., McGehee, D.V., & Manakkal,
N. (1997). Effects of age, system experience, and navigation technique on driving with
an advanced traveler information system. *Human Factors,* 39, 177–179.

Dingus, T.A., Klauer, S.G., Neale, V.L., Petersen, A., Lee, S.E., Sudweeks, J., Perez, M.A.,
Hankey, J., Ramsey, D., Gupta, S., Bucher, C., Doerzaph, Z.R., Jermeland, J., &
Knipling, R.R. (1996). *The 100-Car Naturalistic Driving Study, Phase II—Results of
the 100-Car Field Experiment.* (Report No. DOT HS 810 593). Washington, DC: U.S.
Department of Transportation.

Dobbs, A., Heller, R., & Schopflocker, D. (1998). A comparative approach to identify unsafe
older drivers. *Accident Analysis & Prevention,* 30, 363–370.

Dobbs, B., Carr, D.B., & Morris, J.C. (2002). Management and assessment of the demented
driver. *The Neurologist,* 8, 61–70.

Dobbs, B.M. (2005). *Medical Conditions and Driving: A Review of the Scientific Literature
(1960–2000).* (Report No. DOT HSW 809 690). Washington, DC: U.S. Department
of Transportation.

Dobbs, B.M. & Dobbs, A.R. (1997). *De-Licensing: Mobility and Related Consequences for the Patient and Family Members.* Paper presented at the Transportation Research Board Seventy-Sixth Annual Meeting, Washington, DC.

Dollinger, A. (2007). *Ancient Egyptian Medicine—In Sickness and in Health: Preventative and Curative Health Care.* URL: http://www.reshafim.org.il/ad/egypt/timelines/topics/medicine.htm.

Domey, R.G., McFarland, R.A., & Chadwick, E. (1960). Dark adaptation of a function of time II: A derivation. *Journal of Gerontology,* 15, 267–279.

Drachman, D.A. & Swearer, J.M. (1993). Driving and Alzheimer's disease: The risk of crashes. *Neurology,* 43, 2448–2456.

Drakopouos, A. & Lyles, R.W. (1997). Driver age as a factor in comprehension of left-turn signals. *Transportation Research Record,* 1573, 76–85.

Drakopouos, A. & Lyles, R.W. (2001). An evaluation of age effects on driver comprehension of flashing traffic signal intersections using multivariate multiple response analysis of variance models. *Journal of Safety Research,* 32, 85–106.

Druid, A. (2002). *Vision Enhance System—Does Display Position Matter?* Master's Thesis. Vargarda, Sweden: Linkoping University.

Dubinski, R.M., Graym, C., Hustead, D., Busenbark, K., Vetere-Overfield, B., Wiltfong, D., Parrish, D., & Roller, W.C. (1991). Driving in Parkinson's disease. *Neurology,* 41, 517–520.

Dubinsky, R.M., Williamson, A., Gray, C.S., & Glatt, S.L. (1992). Driving in Alzheimer's disease. *Journal of the American Geriatrics Society,* 40, 1112–1116.

Duchek, J.M., Carr, D.B., Hunt, L., Roe, C.M., Xiong, C., Shah, K.S., & Morris, J.C. (2003). Longitudinal driving performance in early-stage dementia of the Alzheimer type. *Journal of the American Geriatrics Society,* 51, 1342–1347.

Duchek, J.M., Hunt, L., Ball, K., Buckles, V., & Morris, J.C. (1998). Attention and driving performance in Alzheimer's disease. *Journal of Gerontology,* 53B, 130–141.

Eadington, D.W. & Frier, B.M. (1989). Type I diabetes and driving experience: An eight-year cohort study. *Diabetic Medicine,* 6, 137–141.

Eberhard, J.W. (1996). Safe mobility for senior citizens. *IATSS Research,* 20, 29–37.

Eby, D.W. (1995). The convicted drunk driver in Michigan: A profile of offenders. *UMTRI Research Review,* 25(5), 1–11.

Eby, D.W. & Kantowitz, B.K. (2006). Human factors and ergonomics in motor vehicle transportation. *Handbook of Human Factors and Ergonomics, 3rd Edition.* Hoboken, NJ: Wiley.

Eby, D.W. & Kostyniuk, L.P. (1998). Maintaining older driver mobility and well-being with traveler information systems. *Transportation Quarterly,* 52, 45–53.

Eby, D.W. & Kostyniuk, L.P. (1999). A statewide analysis of child safety seat use and misuse in Michigan. *Accident Analysis & Prevention,* 31, 555–566.

Eby, D.W., Kostyniuk, L.P., & Vivoda, J.M. (2001). Restraint use patterns for older child passengers in Michigan. *Accident Analysis & Prevention,* 33, 235–242.

Eby, D.W. & Molnar, L.J. (1999). *Guidelines for Developing Information Systems for the Driving Tourist.* (Report No. ITS-RCE-939430). Ann Arbor, MI: The University of Michigan Intelligent Transportation System Research Center of Excellence.

Eby, D.W. & Molnar, L.J. (2001). Older drivers: Validating a self-assessment instrument with clinical measures and actual driving. *Gerontology,* 41, 370.

Eby, D.W. & Molnar, L.J. (2006). Self-screening by older drivers. *Public Policy and Aging Report,* 15(2), 18–20.

Eby, D.W. & Molnar, L.J. (2008). *Proceedings of the North American License Policies Workshop.* D.W. Eby & L.J. Molnar (Eds.). Washington, DC: AAA Foundation for Traffic Safety.

Eby, D.W., Molnar, L., & Blatt, J. (2005). Common health concerns and symptoms associated with aging: A basis for self-screening by older drivers. *The Gerontologist, Special Issue*, 45, 466.

Eby, D.W., Molnar, L.J., & Kartje, P. (2007). SAFER Driving: Self-screening based on health concerns. *The Gerontologist, Special Issue*, 47, 86.

Eby, D.W., Molnar, L.J., Kartje, P., St Louis, R., Parow, J.E., Vivoda, J.M., & Neumeyer, A. (in press). *Older Adult Self-Screening Based on Health Concerns*. Washington, DC: U.S. Department of Transportation.

Eby, D.W., Molnar, L.J., Kostyniuk, L.P., & Shope, J.T. (1999). The perceived role of the family in older driver reduction and cessation of driving. In *43rd Annual Proceedings: Association for the Advancement of Automotive Medicine*. (pp. 458–460). Des Plaines, IL: Association for the Advancement of Automotive Medicine.

Eby, D.W., Molnar, L.J., & Olk, M.L. (2000). Trends in driver and front-right passenger safety belt use in Michigan: 1984 to 1998. *Accident Analysis & Prevention*, 32, 837–843.

Eby, D.W., Molnar, L.J., & Pellerito, J.M. (2006). Driving cessation and alternative community mobility. In J.M. Pellerito (Ed.), *Driving Rehabilitation and Community Mobility: Principles and Practices*. St. Louis, MO: Elsevier/Mosby, Inc.

Eby, D.W., Molnar, L.J., Shope, J.T., Vivoda, J.M., & Fordyce, T.A. (2003). Improving older driver knowledge and awareness through self-assessment: The *Driving Decisions Workbook*. *Journal of Safety Research*, 34, 371–381.

Eby, D.W., Shope, J.T., Molnar, L.J., Vivoda, J.M., & Fordyce, T.A. (2000). *Improvement of Older Driver Safety Through Self-Evaluation: The Development of a Self-Evaluation Instrument*. (Report No. UMTRI-2000-04). Ann Arbor, MI: University of Michigan Transportation Research Institute.

Eby, D.W., Trombley, D., Molnar, L.J., & Shope, J.T. (1998). *The Assessment of Older Driver's Capabilities: A Review of the Literature*. (Report No. UMTRI-98-24). Ann Arbor, MI: University of Michigan Transportation Research Institute.

Edie, L.C. & Foot, R.S. (1960). Effect of shock waves on tunnel traffic flow. *Highway Research Board Proceedings*, 39, 492–505.

Edwards, C.J., Creaser, J.I., Caird, J.K., Lamsdale, A.M., & Chisholm, S.L. (2002). Older and younger driver performance at complex intersections: Implications for using perception-response time and driving simulation. *Proceedings of the Second International Driving Symposium on Human Factors in Driving Assessment, Training, and Vehicle Design*. Ames, IA: University of Iowa.

Edwards, J.D., Ross, L.A., Ackerman, M.L., Small, B.J., Ball, K.K., Bradley, S., Dodson, J.E. (2008). Longitudinal predictors of driving cessation among older adults from the ACTIVE clinical trial. *Journal of Gerontology: Psychological Sciences*, 63B, P6–P12.

Eisenberg, D.M., Davis, R.B., Ettner, S.L., Appel, S., Wilkey, S., Rompay, M.V., & Kessler, R.C. (1998). Trends in alternative medicine use in the United States, 1990–1997: Results of a follow-up national study. *Journal of the American Medical Association*, 280, 1569–1575.

Eisenberg, D.M., Kessler, R.C., Foster, C., Norlock, F.E., Calkins, D.R., & Delbanco, T.L. (1993). Unconventional medicine in the United States—Prevalence, costs, and patterns of use. *The New England Journal of Medicine*, 328, 246–252.

Eisenhandler, S.A. (1990). The asphalt identikit: Old age and the driver's license. *International Journal of Aging and Human Development*, 30, 1–14.

Eisner, A., Fleming, S.A., Klein, M.L., & Mauldin, W.M. (1987). Sensitivity in older eyes with good acuity: Cross-sectional norms. *Investigating Ophthalmology & Visual Science*, 28, 1824–1831.

Elvik, R. (2003). Effects of road safety of converting intersections to roundabouts: Review of evidence from non-US studies. *Transportation Research Record*, 1847, 1–10.

EndocrineWeb. (2008). *Hypothyroidism: Too Little Thyroid Hormone*. URL: http//www.endo-crineweb.com.

Engleman, G., Asgari-Jirhandeh, N., McLeod, A., Ramsay, C.F., Deary, I.J., & Douglas, N.J. (1996). Self-reported use of CPAP and benefits of CPAP therapy. *Chest*, 109, 1470–1476.

Epilepsy Foundation. (2008). *Frequently Asked Questions*. URL http://www.epilepsyfounda-tion.org

Epstein, A.E., Miles, W.M., Benditt, D.G., Camm, A.J., Darling, E.J., Friedman, P.L., Garson, A., Harvey, J.C., Kidwell, G.A., Klein, G.J., Levine, P.A., Marchlinski, F.E., Prystowsky, E.N., & Wilkoff, B.L. (1996). Personal and public safety issues related to arrhythmias that may affect consciousness: Implications for regulation and physician recommenda-tions. *Circulation*, 94, 1147–1166.

Eriksen, C.W., Hamlin, R.M., & Daye, C. (1973). Aging adults and rate of memory scan. *Bulletin of the Psychonomic Society*, 1, 259–260.

Ervin, R.D., Sayer, J., LeBlanc, D., Bogard, S., Mefford, M., Hagan, M., Bareket, Z., & Winkler, C. (2005). *Automotive Collision Avoidance System Field Operational Test Report: Methodology and Results*. (Report No. DOT HS 809 900). Washington, DC: U.S. Department of Transportation.

European Road Safety Observatory. (2006). *Older Drivers*. URL: http://www.erso.eu.

Evans, D.A., Funkenstein, H.H., Albert, M.S., Scherr, P.A., Cook, N.R., Chown, M.J., Hebert, L.E., Hennekens, C.H., & Taylor, J.O. (1989). Prevalence of Alzheimer's disease in a community population of older persons: Higher than previously reported. *Journal of the American Medical Association*, 262, 2551–2556.

Evans, L. (1991). *Traffic Safety and the Driver*. New York, NY: Van Nostrand Reinhold.

Fancher, P., Ervin, R., Sayer, J., Hagan, M., Bogard, S., Bareket, Z., Mefford, M., & Haugen, J. (1998). *Intelligent Cruise Control Field Operational Test*. (Report No. UMTRI-98-17). Ann Arbor, MI: University of Michigan Transportation Research Institute.

Federal Highway Administration. (1997). *1995 Nationwide Personal Transportation Survey Data Files*. (Report No. FHWA PL-97-034). Washington DC: U.S. Department of Transportation.

Federal Highway Administration. (2002). *Intelligent Transportation Systems in Work Zones: A Cross-Cutting Study*. (Report No. FHWA-OP-02025). Washington, DC: Federal Highway Administration.

Federal Highway Administration. (2003). *Travel Better, Travel Longer: A Pocket Guide to Improve Traffic Control and Mobility for Our Older Population*. Washington, DC: Federal Highway Administration.

Federal Highway Administration. (2007a). *Manual on Uniform Traffic Control Devices: Overview*. URL: http://mutcd.fhwa.dot.gov/kno-overview.htm. Last modified December 5, 2007.

Federal Highway Administration. (2007b). *Intersection: Intersection Safety Facts & Statistics*. URL: http://safety.fhwa.dot.gov/intersections/inter_facts.htm.

Federal Highway Administration. (2008a). Highway Statistics Publications. http://www.fhwa.dot.gov/policy/ohpi/hss/hsspubs.htm. Washington, DC: FHWA, Office of Highway Policy Information. January 16, 2008.

Federal Highway Administration. (2008b). Celebrating 50 Years: The Eisenhower Interstate Highway System. URL: http://www.fhwa.dot.gov/interstate/homepage.cfm.

Fender, T., Wilber, S.T., Stiffler, K.A., & Blanda, M. (2007). Use of ADReS screening tool in an ED population of older drivers. *Annals of Emergency Medicine*, 50, S94.

Fenell, D. (2004). Determinants of supplement use. *Preventive Medicine*, 39, 932–939.

Fiatarone, M.A., Marks, E.C., Ryan, N.D., Meredith, C.N., Lipsitz, L.A., & Evans, W.J. (1990). High intensity strength training in nonagenarians: Effects on skeletal muscles. *Journal of the American Medical Association*, 263, 3029–3034.

Fildes, B. (1997). *Safety of Older Drivers: Strategy for Future Research and Action Initiatives.* (Report No. 118). Victoria, Australia: Monash Accident Research Centre.

Findley, L.J., Barth, J.T., Powers, D.C., Wilhoit, S.C., Boyd, D.G., & Suratt, P.M. (1986). Cognitive impairment in patients with obstructive sleep apnea and associated hypoxemia. *Chest*, 90, 686–690.

Findley, L.J., Fabrizio, M.J., Knight, H., Norcross, B.B., LaForte, A.J., & Suratt, P.M. (1989). Driving simulator performance in patients with sleep apnea. *American Review of Respiratory Diseases*, 140, 529–530.

Findley, L.J., Unverzagt, M.E., & Suratt, P.M. (1988). Automobile accidents involving patients with obstructive sleep apnea. *American Review of Respiratory Diseases*, 138, 337–340.

Finn, J. (2006). Fitting cars to their older drivers. *Healthword: Putting Health Promotion Back in Motion, June 2006*. Washington, DC: American Society on Aging.

Fishbain, D.A., Cutler, R.B., Rosomoff, H.L., & Rosomoff, R.S. (2003). Are opiod-depend/ tolerant patients impaired in driving-related skills? A structured evidence-based review. *Journal of Pain and Symptom Management*, 25, 559–577.

Fisk, G.D., Owsley, C., & Pulley, L.V. (1997). Driving after stroke: Driving exposure, advice, and evaluation. *Archives of Physical Medical Rehabilitation*, 78, 1338–1344.

Fitten, L.J., Perryman, K.M., Wilkinson, R.J., Little, M.M., Pachana, J.R., Mervis, R., Malmgren, R., Siembieda, D.W., & Ganzell, S. (1995). Alzheimer and vascular dementias and driving: A prospective road and laboratory study. *Journal of the American Medical Association*, 273, 1360.

Fjerdingen, L., Jenssen, G.D., Lervag, L.E., van Rijn, L.J., Vaa, T., Kooijman, A., et al. (2004). Report of workshop 1: Vision and perceptual deficiencies as a risk factor in traffic safety. *IMMORTAL workshop on "Vision and perceptual deficiencies as a risk factor in traffic safety,"* Trondheim, Norway, May 9, 2003. Leeds, UK: University of Leeds.

Flannery, A. (2001). Geometric design and safety aspects of roundabouts. *Transportation Research Record*, 1751, 76–81.

Flannery, A. & Datta, T.K. (1996). Modern roundabouts and traffic crash experience in United States. *Transportation Research Record*, 1553, 103–109.

Fogoros, R.N., Elson, J.J., & Bonnet, C.A. (1989). Actuarial incidence and pattern of occurrence of shocks following implantation of automatic implantable cardioverter defibrillator. *Pacing Clinical Electrophysiology*, 12, 1465–1473.

Foley, D. J., Heimovitz, H. K., Guralnik, J., & Brock, D. B. (2002). Driving life expectancy of persons aged 70 years and older in the United States. *American Journal of Public Health*, 92, 1284–1289.

Foley, D.J., Wallace, R.B., & Eberhard, J. (1995). Risk factors for motor vehicle crashes among older drivers in a rural community. *Journal of the American Geriatrics Society*, 43, 776–781.

Fonda, S.J., Wallace, R.B., & Herzog, A.R. (2001). Changes in driving patterns and worsening depressive symptoms among older adults. *Journal of Gerontology Series B: Psychological Sciences and Social Sciences*, 56, S343–S351.

Forbes, T.W. & Holmes, R.S. (1936). Legibility distance of highway destination signs in relation to letter height, letter width and reflectorization. *Proceedings of the Highway Research Board*, 19, 321–335.

Fox, G.K., Bowden, S.C., Bashford, G.M., & Smith, D.S. (1997). Alzheimer's disease and driving: Prediction and assessment of driving performance. *Journal of the American Geriatric Society*, 45, 949–953.

Fox, G.K., Bowden, S.C., & Smith, D.S. (1998). On-road assessment of driving competence after brain impairment: Review of current practice and recommendations for a standardized examination. *Archives of Physical and Medical Rehabilitation*, 79, 1288–1296.

Freeman, E.E., Gange, S.J., Muñoz, B., & West, S.K. (2006). Driving status and risk of entry into long-term care for older adults. *American Journal of Public Health*, 96, 1254–1259.

Freeman, E.E., Muñoz, B., Turano, K.A., & West, S.K. (2006). Measures of visual function and their association with driving modification in older adults. *Investigative Ophthalmology & Visual Science*, 47, 514–520.

French, D.J., West, R.J., Elander, J., & Wilding, J.M. (1993). Decision-making style, driving style, and self-reported involvement in road traffic accidents. *Ergonomics*, 36, 627–644.

Freund, B., Colgrove, L.A., Burke, B.L., & McLeod, R. (2005). Self-rated driving performance among elderly drivers referred for driving evaluation. *Accident Analysis & Prevention*, 37, 613–618.

Freund, B., Gravenstein, S., Ferris, R., Burke, B.L., & Shaheen, E. (2005). Drawing clocks and driving cars: Use of brief tests of cognition to screen driving competency in older adults. *Journal of General Internal Medicine*, 20, 240–244.

Freund, K. (1996). *Transportation Alternatives.* Paper presented at the TRB Mid-Year Meeting on the Safe Mobility of Older Persons, Washington, DC.

Freytag, E. & Sachs, J.C. (1969). Abnormalities of the central vision pathways contributing to Maryland's traffic deaths. *Journal of the American Medical Association*, 204, 119.

Frier, B.M., Matthews, D.M., Steel, J.M., & Duncan, L.J. (1980). Driving and insulin dependent diabetes. *The Lancet*, 315, 1232–1234.

Frucht, S., Rodgers, J.D., Greene, P.E., Gordon, M.F., & Fahn, S. (1999). Falling asleep at the wheel: Motor vehicle mishaps in persons taking pramipexole and ropinrole. *Neurology*, 52, 1908–1910.

Fuller, R. (1984). A conceptualization of driving behavior as threat avoidance. *Ergonomics*, 27, 1139–1155.

Gallo, J.J., Rebok, G.W., & Lesiker, S.E. (1999). The driving habits of adults aged 60 years and older. *Journal of the American Geriatrics Society*, 47, 335–341.

Galski, T., Bruno, T.L., & Ehle, H.T. (1992). Driving after cerebral damage: A model with implications for evaluation. *American Journal of Occupational Therapy*, 46, 324–332.

Garber, N.J. & Srinivasan, R. (1990). Characteristics of accidents involving elderly drivers at intersections. *Transportation Research Record*, 1325, 8–16.

Gaylord, S.A. & Marsh, G.R. (1975). Age differences in the speed of a spatial cognitive process. *Journal of Gerontology*, 30, 674–678.

Gengo, F., Gabos, C., & Miller, J.K. (1989). The pharmacodynamics of diphenhydramine-induced drowsiness and changes in mental performance. *Clinical Pharmacology and Therapeutics*, 45, 15–21.

George, C.F.P., Boudreau, A.C., & Smiley, A. (1996). Simulated driving performance in patients with obstructive sleep apnea. *American Journal of Respiratory and Critical Care Medicine*, 154, 175–181.

George, C.F.P., Nickerson, P.W., Hanly, P.J., Hanly, T.W., & Kyger, M.H. (1987). Sleep apnoea patients have more automobile accidents. *Lancet*, 330, 447.

George, C.F.P. & Smiley, A. (1999). Sleep apnea and automobile crashes. *Sleep*, 22, 790–795.

Gilmore, G.C., Wenk, H., Naylor, L.A., & Stuve, T.A. (1992). Motion perception and aging. *Psychology & Aging*, 7, 654–660.

Gish, K.W., Shoulson, M., & Perel, M. (2002). *Driver Behavior and Performance Using an Infrared Night Vision Enhancement System.* Paper presented at the 81st Annual Meeting of the Transportation Research Board. Washington, DC: Transportation Research Board.

Gonzalez-Rothi, R., Foresman, G., & Block, A. (1988). Do patients with sleep apnea die in their sleep? *Chest*, 94, 531–538.

Goode, K.T., Ball, K., Sloane, M., Roenker, D.L., Roth, D.L., Myers, R.S., & Owsley, C. (2004). Useful field of view and other neurocognitive indicators of crash risk in older adults. *Journal of Clinical Psychology in Medical Settings*, 5, 425–440.

Goodman, M., Bents, F., Tijerina, L., Wierwille, W., Lerner, N., & Benel, D. (1997). *An Investigation of the Safety Implications of Wireless Communications in Vehicles*. (Report No. DOT-HS-808-635). Washington, DC: U.S. Department of Transportation.

Goodwill, C.J. (1984). Mobility for the disabled patient. *International Rehabilitation Medicine*, 6, 1.

Grabowski, D.C., Campbell, C.M., and Morrisey, M.A. (2004). Elderly licensure laws and motor vehicle fatalities. *Journal of the American Medical Association*, 291, 2840–2846.

Graham, J.R., Harrold, J.K., & King, L.E. (1996). Pavement marking retroreflectivity requirements for older drivers. *Transportation Research Record*, 1529, 65–70.

Graveling, A.J., Warren, R.E., & Frier, B.M. (2004). Hypoglycemia and driving in people with insulin-treated diabetes: Adherence to recommendations for avoidance. *Diabetic Medicine*, 21, 1014–1019.

Greenberg, G., Watson, R., & Depula, D. (1987). Neuropsychological dysfunction in sleep apnea. *Sleep*, 10, 254–262.

Greenburg, R.L. (1959). Depositors for confectionary machines. U.S. Patent Office. National Equipment Corporation. New York, NY.

Gregersen, N.P. & Berg, H.Y. (1994). Lifestyle and accidents among young drivers. *Accident Analysis & Prevention*, 26, 297–303.

Grimm, W., Flores, B.T., & Marchlinski, E.E. (1993). Shock occurrences and survival in 241 patients with implantable cardio-defibrillator therapy. *Circulation*, 87, 1880–1888.

Guerrier, J.H., Manivannan, P., & Nair, S.N. (1999). The role of working memory, field dependence, visual search, and reaction time in the left turn performance of older female drivers. *Applied Ergonomics*, 30, 109–119.

Guilleminault, C., Van den Hoed, J., & Mitler, M. (1978). Clinical overview of the sleep apnea syndrome. In C. Guilleminault & W. Dement (Eds.), *Sleep Apnea Syndrome*. New York, NY: AR Liss.

Gurwitz, J.H. (2004). Polypharmacy: A new paradigm or quality drug therapy in the elderly? *Archives of Internal Medicine*, 164, 1957–1959.

Gurwitz, J.H., Field, T.S., Harrold, L.R., Rothschild, J., Debellis, K., Seger, A.C., Cadoret, C., Fish, L.S., Garber, L., Kelleher, M., & Bates, D.W. (2003). Incidence and preventability of adverse drug events among older persons in the ambulatory setting. *Journal of the American Medical Association*, 289, 1107–1116.

Gutierrez, P., Wilson, M.R., & Johnson, C. (1997). Influence of glaucomatous visual field loss on health-related quality of life. *Archives of Ophthalmology*, 115, 777–784.

Haase, G.R. (1977). Diseases presenting as dementia. In C.E. Wells (Ed.), *Dementia, 2nd Edition*. Philadelphia, PA: F.A. Davis.

Hajjar, I., Kotchen, J.M., & Kotchen, T.A. (2006). Hypertension: Trends in prevalence, incidence, and control. *Annual Review of Public Health*, 27, 465–490.

Hakamies-Blomqvist, L. (2004). Safety of older persons in traffic. In *Transportation in an Aging Society: A Decade of Experience*. Washington, DC: Transportation Research Board.

Hakamies-Blomqvist, L. (2006). Are there safe and unsafe drivers? *Transportation Research Part F*, 9, 347–352.

Hakamies-Blomqvist, L., Johansson, K., & Lundberg, C. (1996). Medical screening of older drivers as a traffic safety measure—A comparative Finnish-Swedish evaluation study. *Journal of the American Geriatric Society*, 44, 650–653.

Hakamies-Blomqvist, L., Raitanen, T., & O'Neill, D. (2002). Driver ageing does not cause higher accident rates per km. *Transportation Research Part F*, 5, 271–274.

Hakamies-Blomqvist, L. & Siren, A. (2003). Deconstructing a gender difference: Driving cessation and personal driving history of older women. *Journal of Safety Research,* 34, 383–388.

Hakamies-Blomqvist, L., Siren, A., & Davidse, R. (2004). *Older Drivers—A Review.* (VTI Report 497A). Linkoping, Sweden: Swedish National Road and Transport Research Institute.

Hakamies-Blomqvist, L. & Wahlström, B. (1998). Why do older drivers give up driving? *Accident Analysis & Prevention,* 30, 305–312.

Hakamies-Blomqvist, L.E. (1993). Fatal accidents of older drivers. *Accident Analysis & Prevention,* 25, 19–27.

Hansotia, P. & Broste, S. (1991). The effect of epilepsy and diabetes mellitus on the risk of automobile accidents. *New England Journal of Medicine,* 324, 22–26.

Haraldsson, P.O., Carenfelt, C., & Tingvall, C. (1992). Sleep apnea syndrome symptoms and automobile driving in a general population. *Journal of Clinical Epidemiology,* 45, 821–825.

Harbutt, P. (2007). *Maintaining Mobility: The Transition From Driver to Non-Driver.* Melbourne, Australia: Department of Infrastructure, Policy and Intergovernmental Relations Division.

Harsch, I.A., Stocker, S., Radespiel-Tröger, M., Hahn, E.G., Konturek, P.C., Ficker, J.H., & Lohman, T. (2002). Traffic hypoglycaemias and accidents in patients with diabetes mellitus treated with different antidiabetic regimens. *Journal of Internal Medicine,* 252, 352–360.

Hartford Financial Services Group. (2007). *We Need to Talk...Family Conversations With Older Drivers.* Southington, CT: The Hartford.

Haselkorn, J.K., Mueller, B.A., & Rivara, F.A. (1998). Characteristics of drivers and driving record after traumatic and nontraumatic brain injury. *Archives of Physical Medical Rehabilitation,* 79, 738–742.

Hatakka, M. (1998). Novice drivers' risk- and self-evaluations [in Finnish]. Turun yliopiston julkaisuja Painosalama Oy, Turku.

Hatakka, M., Keskinen, E., Gregersen, N.P., Glad, A., & Hernetkoski, K. (2002). From control of the vehicle to personal self-control: Broadening the perspectives to driver education. *Transportation Research Part F,* 5, 201–215.

Hauer, E. (1988). The safety of older persons at intersections. *Transportation in an Aging Society: Improving Mobility and Safety for Older Persons—Volume 2: Special Report 218.* Washington, DC: Transportation Research Board.

Hauser, W.A. & Kurland, L.T. (1975). The epidemiology of epilepsy in Rochester, Minnesota, 1935 through 1967. *Epilepsia,* 16, 1–66.

Hawkins, H.G. (1992). Evolution of the MUTCD: Early standards for traffic control devices. *ITE Journal,* July, 23–26.

Hawkins, H.G., Womack, K.N., & Mounce, J.M. (1993). Driver comprehension of regulatory signs, warning signs, and pavement markings. *Transportation Research Record,* 1403, 67–82.

Haymes, S.A., LeBlanc, R.P., Nicolela, M.T., Chiasson, L.A., & Chauhan, B.C. (2007). Risk of falls and motor vehicle collisions in glaucoma. *Investigative Ophthalmology & Visual Science,* 48, 1149–1155.

Heglin, H. (1956). Problem solving set in different age groups. *Journal of Gerontology,* 11, 310–317.

Heikkilä, V-M., Turkka, A., Korpelainen, J., Kallanranta, T., & Summala, H. (1998). Decreased driving ability in people with Parkinson's disease. *Journal of Neurology, Neurosurgery, and Psychiatry,* 64, 325–330.

Helling, D.K., Lemke, J.H., Semla, T.P., Wallace, R.B., Lipson, D.P., & Cornoni-Huntley, J. (1987). Medication use in the elderly: The Iowa 65+ Rural Health Study. *Journal of the American Geriatrics Society,* 35, 4–12.

Hemmelgarn, B., Suissa, S., Huang, A., Boivin, J.F., & Pinard, G. (1997). Benzodiazepine use and the risk of motor vehicle crash in the elderly. *Journal of the American Medical Association*, 278, 27–31.

Henderson, S. & Suen, S.L. (1999). Intelligent transportation systems: A two-edged sword for older drivers? *Transportation Research Record*, 1679, 58–63.

Hennessey, D.F. (1995). *Vision Testing of Renewal Applicants Crashes Predicted When Compensation for Impairment Is Inadequate.* (Report RSS-95-152). Sacramento, CA: California Department of Motor Vehicles.

Hennessey, D.F. (2003). *Three-Tier Assessment System: Concept and Study.* Unpublished report.

Herman, J.F. & Bruce, P.R. (1983). Adults' mental rotation of spatial information: Effects of age, sex, and cerebral laterality. *Experimental Aging Research,* 9, 83–85.

Heron, A. & Chown, S.M. (1967). *Age and Function.* London, UK: Churchill.

Herriotts, P. (2005). Identification of vehicle design requirement for older drivers. *Applied Ergonomics,* 36, 255–262.

Hills, B.L. (1975). *Some Studies of Movement Perception, Age and Accidents* (Report No. TTRL-137UC). Crowthorne, UK: Transport and Road Research Laboratory.

Hills, B.L. (1980). Vision, visibility, and perception in driving. *Perception,* 9, 183–216.

Hindmarch, I. (1988). The psychopharmacological approach: Effects of psychotropic drugs on car handling. *International Clinical Psychopharmacology,* 3, 74–79.

Hobson, D.E., Lang, A.E., Martin, W.R.W., Razmy, A., Rivest, J., & Fleming, J. (2002). Excessive daytime sleepiness and sudden-onset sleep in Parkinson disease. *Journal of the American Medical Association,* 287, 455–463.

Hoedemaeker, M. & Brookhuis, K.A. (1998). Behavioural adaptation to driving with an adaptive cruise control (ACC). *Transportation Research Part F,* 1, 95–106.

Hoffman, C.S., Price, A.C., Garrett, E.S., & Rothstein, W. (1959). Effect of age and brain damage on depth perception, *Perceptual Motor Skills,* 9, 283–286.

Hofstetter, H.W. (1965). A longitudinal study of amplitude changes in presbyopia. *American Journal of Optometry,* 42, 3–8.

Hofstetter, H.W. & Bertsch, J.D. (1976). Does stereopsis change with age? *American Journal of Ophthalmology & Physiological Optics,* 53, 664–667.

Hogan, D.B. (2005). Which older patients are competent to drive? Approaches to office-based assessment. *Canadian Family Physician,* 51, 362–368.

Holland, C.A. & Rabbitt, P.M.A. (1991). Aging memory: Use versus impairment. *British Journal of Psychology,* 82, 29–38.

Hood, D.C. & Finkelstein, M.A. (1986). Sensitivity to light. In K.R. Boff, L. Kaufman, & J.P. Thomas (Eds.), *Handbook of Perception and Human Performance: Volume 1, Sensory Processes and Perception.* New York, NY: John Wiley and Sons.

Horberry, T., Anderson, J., & Regan, M.A. (2006). The possible safety benefits of enhanced road markings: A driving simulator evaluation. *Transportation Research Part F,* 9, 77–87.

Horstman, S., Hess, C.W., Bassetti, C., Gugger, M., & Mathias, J. (2000). Sleepiness-related accidents in sleep apnea patients. *Sleep,* 23, 1–7.

Hoyer, W.J. & Familant, M.E. (1987). Adult age differences in the rate of processing expectancy information. *Cognitive Development,* 2, 59–70.

Hu, P.S. & Reuscher, T.R. (2004). *Summary of Travel Trends: 2001 National Household Survey.* Washington, DC: U.S. Department of Transportation.

Huaman, A.G. & Sharpe, J.A. (1993). Vertical saccades in senescence. *Investigative Ophthalmology & Visual Science,* 34, 2588–2595.

Hughes, C.P., Berg, L., Danziger, W.L., Coben, L.A., & Martin, R.L. (1982). A new clinical scale for the staging of dementia. *British Journal of Psychiatry,* 140, 556–572.

Hultsch, D.F., Hammer, M.,& Small, B.J. (1993). Age differences in cognitive performance in later life: Relationships to self-reported health and activity style. *Journal of Gerontology: Psychological Sciences*, 48, 1–11.

Hultsch, D.F., Hertzog, C., Small, B.J., & Dixon, R.A. (1999). Use it or lose it: Engaged lifestyle as a buffer of cognitive decline in aging? *Psychology and Aging*, 14, 245–263.

Hummer, J.E., Montgomery, R.E., & Sinha, K.C. (1990). Motorists understanding of and preferences for left-turn signals. *Transportation Research Record*, 1281, 136–147.

Hunt, L.A. (1996). A profile of occupational therapists as driving instructors and evaluators for elderly drivers. *International Association of Traffic Safety and Science Research*, 20, 46–47.

Hunt, L.A. & Arbesman, M. (2008). Evidence-based and occupational perspective of effective interventions for older clients that remediate or support improved driving performance. *The American Journal of Occupational Therapy*, 62, 136–148.

Hunt, L.A., Morris, J.C., Edwards, D., & Wilson, B.S. (1993). Driving performance in persons with mild senile dementia of the Alzheimer's type. *Journal of the American Geriatric Society*, 41, 747–753.

Hunt, L.A., Murphy, C., Carr, D.B., Duchek, J.M., Buckles, V., & Morris, J.C. (1997). Reliability of the Washington University Road Test: A performance-based assessment for drivers with dementia of the Alzheimer's type. *Archives of Neurology*, 54, 707–712.

Hunter-Zaworski, K.M. (1990). T-intersection simulator performance of drivers with physical limitations. *Transportation Research Record*, 1281, 11–15.

Insurance Institute for Highway Safety. (2004). *U.S. Driver Licensing Renewal Procedures for Older Drivers as of May 2004*. Available at http://www.hwysafety.org/safety%5Ffacts/ state%5Flaws/older%5Fdrivers.htm. Accessed April 12, 2005.

Insurance Institute for Highway Safety. (2007). *Fatality Facts: Older People*. http://www.iihs. org/research/qanda/older_people.html. August, 2007.

Insurance Institute for Highway Safety. (2008). *Fatality Facts 2006: Older People*. URL: http://www.iihs.org/research/fatality_facts_2006/olderpeople.html.

Internet Drug News Inc. (2007). *Antidepressant Drug Database: Information on Antidepressants Used to Treat Clinical Depression*. URL: http://www.coreynahman.com/antidepressant-drugsdatabase.html.

Ivers, R.Q., Mitchell, P., & Cumming, R.G. (1999). Sensory impairment and driving: the Blue Mountains Eye Study. *American Journal of Public Health*, 89, 85–87.

Jacewicz, M.M. & Hartley, A.A. (1979). Rotation of mental images by young and old college students: Effects of familiarity. *Journal of Gerontology*, 34, 396–403.

Jackson, G.R., Owsley, C., & Cideciyan, A.V. (1997). Rod vs. cone sensitivity loss in age-related macular degeneration. *Investigative Ophthalmology & Visual Science*, 38, S354.

Jacobs, R.I., Johnson, A.W., & Cole, B.L. (1975). The visibility of alphabetic and symbol traffic signs. *Australian Road Research*, 5, 68–86.

James, C.L., Wochinger, K., James, W.S., & Boehm-Davis, D. (1997). Visual, perceptual, and cognitive measures as predictors of collision detection in older drivers. *Proceedings of the Human Factors and Ergonomics Society 41st Annual Meeting*. Santa Monica, CA: Human Factors and Ergonomics Society.

Jani, S.N. (1966). The age factor in stereopsis screening. *American Journal of Ophthalmology*, 43, 653–657.

Janke, M. (1991). Accidents, mileage and the exaggeration of risk. *Accident Analysis & Prevention*, 23, 183–188.

Janke, M. & Eberhard, J. (1998). Assessing medically impaired older drivers in a licensing agency setting. *Accident Analysis & Prevention*, 30, 347–361.

Janke, M.K. (1994). *Age-Related Disabilities That May Impair Driving and Their Assessment: Literature Review*. (Report No. RSS-94-156). Sacramento, CA: California Department of Motor Vehicles.

Janke, M.K. (2001). Assessing older drivers—Two studies. *Journal of Safety Research*, 32, 43–74.

Jessor, R. (1987). Problem-behavior theory, psychosocial development, and adolescent problem drinking. *Addiction*, 82, 331–342.

Johansson, K., Bronge, L., Lundberg, C., Persson, A., Seideman, M., & Viitanen, M. (1996). Can a physician recognize an older driver with increased crash risk potential? *Journal of the American Geriatrics Society*, 44, 1198–1204.

Johnson, C.A. & Keltner, J.L. (1983). Incidence of visual field loss in 20,000 eyes and its relationship to driving performance. *Archives of Ophthalmology*, 3, 371–375.

Johnson, J.E. (1995). Rural elders and the decision to stop driving. *Journal of Community Health Nursing*, 12, 131–138.

Johnson, J.E. (1999). Urban older adults and the forfeiture of a driver's license. *Journal of Gerontological Nursing*, 25, 12–18.

Jones, J.G., McCann, J., & Lassere, M.N. (1991). Driving and arthritis. *British Journal of Rheumatology*, 30, 361–364.

Jones, P. (2004). *All About Multiple Sclerosis*. URL: http://www.mult-sclerosis.org.

Jones, R.K., Shinar, D., & Walsh, J.M. (2003). *State of Knowledge of Drug-Impaired Driving*. (Report No. DOT HS 809 642). Washington, DC: U.S. Department of Transportation.

Jordan, P.W. (1985). Pedestrians and cyclists at roundabouts. *Proceedings of the 3rd National Local Government Engineers Conference*. Adelaide, Australia.

Kahneman, D. (1973). *Attention and Effort*. Englewood Cliffs, NJ: Prentice-Hall.

Kales, A., Caldwell, A.B., Cadieux, R.J., Vela-Bueno, A., Ruch, L.G., & Mayes, S.D. (1985). Severe obstructive sleep apnea-II: Associated psychopathology and psychosocial consequences. *Journal of Chronic Diseases*, 38, 427–434.

Kanianthra, J., Carter, A., & Preziotti, G. (2000). *Field Operational Test Results of an Automated Collision Notification System*. (Report No. 2000-01-C041). Warendale, PA: SAE International.

Kannel, W.B., Gagnon, D.R., & Cupples, L.A. (1990). Epidemiology of sudden coronary death: Population at risk. *Canadian Journal of Cardiology*, 6, 439–444.

Kantor, B., Mauger, L., Richardson, V.E., & Unroe, K.T. (2004). An analysis of an older driver evaluation program. *Journal of Gerontology*, 52, 1326–1330.

Kaplan, G.A. (1995). Where do shared pathways lead? Some reflections on a research agenda. *Psychosomatic Medicine*, 57, 208–212.

Kapoor, W.N. (1994). Syncope in older persons. *Journal of the American Geriatric Society*, 42, 426–436.

Kapoor, W.N., Hammill, S.C., & Gersh, B.J. (1989). Diagnosis and natural history of syncope and the role of invasive electrophysiologic testing. *American Journal of Cardiology*, 63, 730–734.

Kapoor, W.N., Karpf, M., Wieand, S., Peterson, J.R., & Levey, G.S. (1983). A prospective evaluation and follow-up of patients with syncope. *New England Journal of Medicine*, 309, 197–204.

Katzman, R. (1987). Alzheimer's disease: Advances and opportunities. *Journal of the American Geriatric Society*, 35, 69–73.

Kausler, D.H. (1990). Automaticity of encoding and episodic memory processes. In E.A. Lovelace (Ed.), *Aging and Cognition: Mental Processes, Self-Awareness, and Interventions*. Amsterdam, The Netherlands: Elsevier.

Kausler, D.H. (1991). *Experimental Psychology, Cognition, and Human Aging*. New York, NY: Springer-Verlag.

Kausler, D.H. & Puckett, J.M. (1980). Adult age differences in recognition memory for a non-semantic attribute. *Experimental Aging Research*, 6, 349–355.

Kausler, D.H. & Puckett, J.M. (1981). Adult age differences in memory for modality attributes. *Experimental Aging Research*, 7, 117–125.

Kay, L., Bundy, A., Clemson, L., & Jolly, N. (2008). Validity and reliability of the on-road driving assessment with senior drivers. *Accident Analysis & Prevention*, 40, 751–759.

Keeney, A.H. (1968). Relationship of ocular pathology and driving impairment. *Transactions of the American Academy of Ophthalmology and Otolaryngology*, 21, 22–27.

Keeney, A.H., Garbey, J.L., & Brunker, G.F. (1981). Current experience with the monocular driver of Kentucky. In *Proceedings of the American Association for Automotive Medicine*. Des Plaines, IL: AAAM.

Kelso, J.A.S. (1982). *Human Motor Behavior: An Introduction*. Hillsdale, NJ: Lawrence Erlbaum Associates.

Kennedy, R.L., Henry, J., Chapmen, A.J., Nayar, R., Grant, P., & Morris, A.D. (2002). Accidents in patients with insulin-treated diabetes: Increased risk of low-impact falls but not motor vehicle crashes—A prospective register-based study. *The Journal of Trauma—Injury, Infection, & Critical Care*, 52, 660–666.

Ker, K., Roberts, I., Collier, T., Beyer, F., Bunn, F., & Frost, C. (2005). Post-license driver education for the prevention of road traffic crashes: A systematic review of randomized controlled trials. *Accident Analysis & Prevention*, 37, 305–313.

Kerschner, H. & Hardin, J. (2006). *Transportation Innovations for Seniors: A Report From Rural America*. Pasadena, CA: Beverly Foundation and the Community Transportation Association of America.

Keskin, E., Ota, H., & Katila, A. (1989). Older drivers fail in intersections: Speed discrepancies between older and younger male drivers. *Accident Analysis & Prevention*, 30, 323–330.

Keskinen, E. (1996). Why do young drivers have more accidents? Junge Fahrer Und Fahrerinnen. Referate der Esten Interdiziplinären Fachkonferenz 12–14 Dezember 1994 in Köln. Berichte der Bundesanstalt fur Strassenwesen. Mensch und Sicherheit, Heft M 52.

Keskinen, E. (2007). What is GDE all about and what it is not. In W. Henriksson, T. Stenlund, A. Sundstrom, & M. Wiberg (Eds.), *Proceedings From The GDE-Model as a Guide in Driver Training and Testing*. Umea, Sweden: Umea University.

Keskinen, E., Hatakka, M., Laapotti, S., Katila, A., & Peraaho, M. (2004). Driver behavior as a hierarchical system. In T. Rothengatter & R.D. Huguenin (Eds.), *Traffic and Transport Psychology: Theory and Application: Proceedings of the ICTTP 2000*. New York, NY: Elsevier.

Kidd, E.A. & Laughery, K.R. (1964). *A Computer Model of Driving Behavior: The Highway Intersection Situation*. (Report No. VJ-1843-V-1). Ithaca, NY: Cornell Aeronautical Laboratory.

Kiefer, R.J., Cassar, M.T., Flannagan, C.A., LeBlanc, D.J., Palmer, M.D., Deering, R.K., & Shulman, M.A. (2003). *Forward Collision Warning Requirements Project: Refining the CAMP Crash Alert Timing Approach by Examining "Last Second" Braking and Lane Change Maneuvers Under Various Kinematic Conditions*. (Report No. DOT HS 809 574). Washington, DC: U.S. Department of Transportation.

Kihl, M., Brennan, D., Gabhawala, N., List, J., & Mittal, P. (2005). *Livable Communities: An Evaluation Guide*. Washington, DC: AARP Public Policy Institute & Arizona State University Herberger Center for Design Excellence.

Klavora, P. & Heslegrave, R.J. (2002). Senior drivers: An overview of problems and intervention strategies. *Journal of Aging and Physical Activity*, 10, 322–335.

Klavora, P., Heslegrave, R.J., & Young, M. (2000). Driving skills in elderly persons with stroke: Comparison of two new assessment options. *Archives of Physical Medical Rehabilitation*, 81, 701–705.

Klebelsberg, D. (1977). Das Modell der subjektiven und objektiven Sicherheit. [In German]. *Schweizerische Zeitschrift für Psychologie und ihre Anwendungen*, 36, 285–294.

Klein, R., Klein, B.E.K., & Cruikshanks, K.J. (1999). The prevalence of age-related macul-opathy by geographic region and ethnicity. *Progress in Retinal and Eye Research*, 18, 371–389.

Kline, D., Schieber, F., Abusamra, L.C., & Coyne, A.C. (1983). Age, the eye, and the visual channels: Contrast sensitivity and response speed. *Journal of Gerontology*, 38, 211–216.

Kline, T.J.B., Ghali, L.M., Kline, D.W., & Brown, S. (1990). Visibility distance of highway signs among young, middle-aged, and elderly observers: Icons are better than text. *Human Factors*, 32, 609–619.

Knecht, J. (1977). The multiple sclerosis patient as a driver [in German]. *Journal Suisse De Medecine*, 107, 373–378.

Knoblauch, R., Nitzburg, M., & Seifert, R. (1997). *An Investigation of Older Driver Freeway Needs and Capabilities*. (Report No. FHWA-RD-95-194). Washington, DC: Federal Highway Administration.

Kochera, A., Straight, A., & Guterbock, T. (2005). *Beyond 50.05: A Report to the Nation on Livable Communities: Creating Environments for Successful Aging*. Washington, DC: AARP.

Koepsell, T., McCloskey, L., Wolf, M., Moudon, A.V., Buchner, D., Kraus, J., & Patterson, M. (2002). Crosswalk marking and the risk of pedestrian-motor vehicle collisions in older pedestrians. *Journal of the American Medical Association*, 288, 2136–2142.

Koppa, R. (2004). Automotive adaptive equipment and vehicle modifications. In *Transportation in an Aging Society: A Decade of Experience*. Washington, DC: Transportation Research Board.

Kostyniuk, L.P., Eby, D.W., Christoff, C., & Hopp, M.L. (1997a). *An Evaluation of Driver Response to the TetraStar Navigation Assistance System by Age and Sex*. (Report No. UMTRI-97-33). Ann Arbor, MI: University of Michigan Transportation Research Institute.

Kostyniuk, L.P., Eby, D.W., Christoff, C., & Hopp, M.L. (1997b). *The FAST-TRAC Natural Use Leased-Car Study: An Evaluation of User Perceptions and Behaviors of Ali-Scout by Age and Gender*. (Report No. UMTRI-97-09). Ann Arbor, MI: University of Michigan Transportation Research Institute.

Kostyniuk, L.P., Eby, D.W., & Miller, L.L. (2003). *Crash Trends of Older Drivers in Michigan: 1998–2002*. (Report No. UMTRI-2003-22). Ann Arbor, MI: University of Michigan Transportation Research Institute.

Kostyniuk, L.P. & Molnar, L.J. (2005). Driving self-restriction among older adults: Health, age, and sex effects. *The Gerontologist: Special Issue*, 45, 143.

Kostyniuk, L.P. & Molnar, L.J. (2007). Self regulation of driving by older women. *Transportation Research Board 86th Annual Meeting Final Program*. Washington DC: Transportation Research Board.

Kostyniuk, L.P. & Molnar, L.J. (in press). Driving self-restriction among older adults: Health, age, and sex effects. *Accident Analysis & Prevention*.

Kostyniuk, L.P. & Shope, J.T. (1998). *Reduction and Cessation of Driving Among Older Drivers: Focus Groups* (Report No. UMTRI-98-26). Ann Arbor, MI: University of Michigan Transportation Research Institute.

Kostyniuk, L.P. & Shope, J.T. (2003). Driving and alternatives: Older drivers in Michigan. *Journal of Safety Research*, 34, 407–414.

Kostyniuk, L.P., Shope, J.T., & Molnar, L.J. (2000). *Reduction and Cessation of Driving Among Older Drivers in Michigan: Final Report* (Report No. UMTRI-2000-06). Ann Arbor, MI: University of Michigan Transportation Research Institute.

Kostyniuk, L.P., Streff, F.M., & Eby, D.W. (1997). *The Older Driver and Navigation Assistance Technologies*. (Report No. UMTRI-97-47). Ann Arbor, MI: University of Michigan Transportation Research Institute.

Kostyniuk, L.P., Trombley, D.A., & Shope, J.T. (1998). *The Process of Reduction and Cessation of Driving Among Older Drivers: A Review of the Literature* (Report. No. UMTRI-98-23). Ann Arbor, MI: University of Michigan Transportation Research Institute.

Kou, W.H., Calkins, H., Lewis, R.R., Bolling, S.F., Kirsch, M.M., Langberg, J.J., de Buitleir, M., Sousa, J., El-Atassi, R., & Morady, F. (1991). Incidence of loss of consciousness during automatic implantable cardioverter-defibrillator shocks. *Annals of Internal Medicine*, 115, 942–945.

Kramer, A.F., Cassavaugh, N., Horrey, W.J., & Mayhugh, J.L. (2007). Influence of age and proximity warning devices on collision avoidance in simulated driving. *Human Factors*, 49, 935–949.

Kua, A., Korner-Bitensky, N., Desrosiers, J., Man-Hong-Sing, M., & Marshall, S. (2007). Older driver retraining: A systematic review of effectiveness. *Journal of Safety Research*, 38, 81–90.

Kurtzke, J.F. (1985). Epidemiology of cerebrovascular disease. In F. McDowell & L.R. Caplan (Eds.), *Cerebrovascular Survey Report for the National Institutes of Neurological and Communicative Disorders and Stroke, Revised*. Rochester, MN: Whiting Press.

Laapotti, S. & Keskinen, E. (2004). Has the difference in accident patterns between male and female drivers changed between 1984 and 2000? *Accident Analysis & Prevention*, 36, 577–584.

Laberge-Nadeau, C., Dionne, G., Ékoé, J., Mamet, P., Desjardins, D., Messier, S., & Maage, U. (2000). Impact of diabetes on crash risk of truck-permit holders and commercial drivers. *Diabetes Care*, 23, 612–617.

Land Transport NZ. (2006). *Supporting Older Drivers: Help Your Older Friend or Relative Stay Mobile—Safely*. Wellington, New Zealand: Land Transport NZ.

Langford, J., Braitman, K., Charlton, J., Eberhard, J., O'Neill, D., Staplin, L., & Stutts, J. (2007). *Communiqué from the Panel*, TRB Workshop 2007, Washington, DC.

Langford, J., Fitzharris, M., Koppel, S., & Newstead, S. (2004). Effectiveness of mandatory license testing for older drivers in reducing crash risk among urban older Australian drivers. *Traffic Injury Prevention*, 5, 326–335.

Langford, J., Fitzharris, M., Newstead, S., & Koppel, S. (2004). Some consequences of different older driver licensing procedures in Australia. *Accident Analysis & Prevention*, 36, 993–1001.

Langford, J. & Koppel, S. (2006a). Epidemiology of older driver crashes—Identifying older driver risk factors and exposure patterns. *Transportation Research Part F*, 9, 309–321.

Langford, J. & Koppel, S. (2006b). The case for and against mandatory age-based assessment of older drivers. *Transportation Research Part F*, 9, 353–362.

Langford, J. & Oxley, J. (2006). Using the safe system approach to keep older drivers safely mobile. *IATSS Research*, 30, 97–109.

Langford, J., Methorst, R., & Hakamies-Blomqvist, L. (2006). Older drivers do not have a high crash risk—A replication of low mileage bias. *Accident Analysis & Prevention*, 38, 574–578.

Langlois, J.A., Keyl, P.M., Guralnik, J.M., Foley, D.J., Marottoli, R.A., & Wallace, R.B. (1997). Characteristics of older pedestrians who have difficulty crossing the street. *American Journal of Public Health*, 87, 393–397.

Lapidot, M.B. (1987). Does the brain age uniformly? Evidence from effects of smooth pursuit eye movements on verbal and visual tasks. *Journal of Gerontology*, 42, 329–331.

Larsen, L. & Kines, P. (2002). Multi-disciplinary in-depth investigations of head on and left turn road collisions. *Accident Analysis & Prevention*, 34, 367–380.

Laughery, K.R., Anderson, T.E., & Kidd, E.A. (1967). A computer simulation model of driver-vehicle performance at intersections. In *Proceedings of the A.C.M. Annual Conference*. Washington, DC.

Laurberg, P., Bülow, P.I., Perersen, K.M., & Vestergaard, H. (1999). Low incidence rate of overt hypothyroidism compared with hyperthyroidism in an area with moderately low iodine intake. *Thyroid*, 9, 33–38.

Laux, L.F. & Brelsford, J., Jr. (1990). *Age-Related Changes in Sensory, Cognitive, Psychomotor and Physical Functioning and Driving Performance in Drivers Aged 40 to 92*. Washington DC: AAA Foundation for Traffic Safety.

LeBlanc, D., Sayer, J., Winkler, C., Ervin, R., Bogard, S., Devonshire, J., Mefford, M., Hagan, M., Bareket, Z., Goodsell, R., & Gordon, T. (2006). *Road Departure Crash Warning System Field Operational Test: Methodology and Results*. (Report No. UMTRI-2006-9-1). Ann Arbor, MI: University of Michigan Transportation Research Institute.

Lee, H.C., Drake, V., & Cameron, D. (2002). Identification of appropriate assessment criteria to measure older adults' driving performance in simulated driving. *Australian Occupational Therapy Journal*, 49, 138–145.

Legh-Smith, J., Wade, D.T., & Hower, R.L. (1986). Driving after stroke. *Journal of the Royal Society of Medicine*, 79, 200–203.

Lesikar, S.E., Gallo, J.J., Rebok, G.W., & Keyl, P.M. (2002). Prospective study of brief neuropsychological measures to assess crash risk in older primary care patients. *The Journal of the American Board of Family Medicine*, 15, 11–19.

Leveille, S.G., Buchner, D.M., Koepsel, T.D., McCloskey, L.W., Wolf, M.E., & Wagner, E.H. (1994). Psychoactive medications and injurious motor vehicle collisions involving older drivers. *Epidemiology*, 5, 591–598.

Levy, D.T., Vernick, J.S., & Howard, K.A. (1995). Relationship between driver's license renewal policies and fatal crashes involving drivers 70 years and older. *Journal of the American Medical Association*, 274, 1026–1030.

Li, G., Braver, E., & Chen, L. (2003). Fragility versus excessive crash involvement as determinants of high death rates per vehicle-mile of travel among older drivers. *Accident Analyses & Prevention*, 35, 227–235.

Li, H., Weitzel, M., Easley, A., Barrington, W., & Windle, J. (2000). Related potential risk of vasovagal syncope for motor vehicle driving. *American Journal of Cardiology*, 85, 184–186.

Liddle, J. & McKenna, K. (2003). Older drivers and driving cessation. *British Journal of Occupational Therapy*, 66, 125–132.

Liddle, J., McKenna, K., & Bartlett, H. (2007). Improving outcomes for older retired drivers: The UQDRIVE program. *Australian Occupational Therapy Journal*, 54, 303–306.

Liddle, J., McKenna, K., & Broome, K. (2004). *Older Road Users: From Driving Cessation to Safe Transportation*. Brisbane, Australia: University of Queensland.

Light, L.L. (1966). Memory and aging. In E.L. Bjork & R.A. Bjork (Eds.), *Memory*. New York, NY: Academic Press.

Light, L.L., La Voie, D., Owens, S.A., & Mead, G. (1992). Direct and indirect measures for memory modality in young and older adults. *Journal of Experimental Psychology: Learning, Memory, & Cognition*, 18, 1284–1297.

Lincoln, N.B. & Radford, K.A. (2008). Cognitive abilities as predictors of safety to driving in people with multiple sclerosis. *Multiple Sclerosis*, 14, 123–128.

Lings, S. & Jensen, P.B. (1991). Driving after stroke: A controlled laboratory investigation. *International Disabilities Studies*, 13, 74–82.

Linnoila, M. & Hakkinen, S. (1973). Effects of diazepam and codeine, alone and in combination with alcohol, on simulated driving. *Clinical Pharmacology and Therapeutics*, 15, 368–373.

Lococo, K.H. & Staplin, L. (2005). *Strategies for Medical Advisory Boards and Licensing Review*. (Report No. DOT HS 809 874). Washington, DC: U.S. Department of Transportation.

Lococo, K.H. & Staplin, L. (2006). *Literature Review of Polypharmacy and Older Drivers: Identifying Strategies to Collect Drug Usage and Driving Functioning Among Older Drivers.* (Report No. DOT HS 810 558). Washington, DC: U.S. Department of Transportation.

Logsdon, R.G., Teri, L., & Larson, E.B. (1992). Driving and Alzheimer's disease. *Journal of General Internal Medicine,* 7, 583–588.

Long, G.M. & Crambert, R.F. (1989). The nature and basis of age-related changes in dynamic visual acuity. *Psychology of Aging,* 5, 138–143.

Lord, D., van Schalkwyk, I., Chrysler, S., & Staplin, L. (2007). A strategy to reduce older driver injuries at intersections using more accommodating roundabout design practices. *Accident Analysis & Prevention,* 39, 427–432.

Lord, D., van Schalkwyk, I., Staplin, L., & Chrysler, S. (2005). *Reducing Older Driver Injuries at Intersections Using More Accommodating Roundabout Design Practices.* (Report No. TTI-CTS-05-01). College Station, TX: Texas Transportation Institute.

Louwerens, J.W., Brookhuis, K.A., & O'Hanlon, J.F. (1984). *The Effects of the Anti-Depressants Oxaprotilene, Mianserin, Amitriptyline, and Doxepin Upon Actual Driving Performance.* Groningen, Netherlands: University of Groningen Traffic Research Center.

Lowenfeld, I.E. (1979). Pupillary changes related to age. In H.S. Thompson (Ed.), *Topics in Neuro-Ophthalmology.* Baltimore, MD: Williams and Wilkins.

Lucas-Blaustein, M.J., Filipp, L., Dungan, C., & Tune, L. (1988). Driving in patients with dementia. *Journal of the American Geriatric Society,* 36, 1087–1092.

Lundberg, C., Caneman, G., Samuelsson, S., Hakamies-Blomqvist, L., & Almkvist, O. (2003). The assessment of fitness to drive after a stroke: The Nordic Stroke Driver Screening Assessment. *Scandinavian Journal of Psychology,* 44, 23–30.

Lundberg, C., Hakamies-Blomqvist, L., Almkvist, O., & Johansson, K. (1998). Impairments of some cognitive functions are common in crash-involved older drivers. *Accident Analysis & Prevention,* 30, 371–377.

Lundqvist, A., Gerdle, B., & Roennberg, J. (2000). Neuropsychological aspects of driving after a stroke—in the simulator and on the road. *Applied Cognitive Psychology,* 14, 135.

MacGregor, J.M., Freeman, J.H., & Zhang, D. (2001). A traffic sign recognition test can discriminate between older drivers who have and have not had a motor vehicle crash. *Journal of the American Geriatrics Society,* 49, 466–469.

MacMahon, M., O'Neill, D., & Kenny, R.A. (1996). Syncope: Driving advice is frequently overlooked. *Postgraduate Medicine Journal,* 72, 561–563.

Madden, D.J. & Greene, H.A. (1987). From retina to response: Contrast sensitivity and memory retrieval during visual word recognition. *Experimental Aging Research,* 13, 15–21.

Malfetti, J.W. (1985). *Needs and Problems of Older Drivers: Survey Results and Recommendations.* Falls Church, VA: AAA Foundation for Traffic Safety.

Maloney, J., Masterson, M., Khoury, D., Trohman, R., Wilkoff, B., Simmons, T., Morant, V., & Castle, L. (1991). Clinical performance of the implantable cardioverter defibrillator: Electrocardiographic documentation of 101 spontaneous discharges. *Pace,* 14, 280–285.

Maltz, M. & Shinar, D. (2004). Imperfect in-vehicle collision avoidance warning systems can aid drivers. *Human Factors,* 46, 357–366.

Mangione, C.M., Ajani, U.A., & Padan, A. (1994). Impact of age-related macular degeneration on visual functioning. *Journal of the American Geriatrics Society,* 42, SA73.

Man-Son-Hing, M., Marshall, S.C., Molnar, F.J., & Wilson, K.G. (2007). Systematic review of driving risk and the efficacy of compensatory strategies in persons with dementia. *Journal of the American Geriatrics Society,* 55, 878–884.

Marcotte, T.D., Rosenthal, T.J., Corey-Bloom, J., Roberts, E., Lampinen, S., & Allen, R.W. (2005). The impact of cognitive deficits and spasticity on driving simulator performance in multiple sclerosis. In *Proceedings of the Third International Driving Symposium on Human Factors in Driver Assessment, Training, and Vehicle Design*. Ames, IA: University of Iowa.

Margolis, K.L., Kerani, R.P., McGovern, P., Songer, T., Cauley, J.A., & Ensrud, K.E. (2002). Risk factors for motor vehicle crashes in older woman. *Journal of Gerontology*, 57A, M186–M191.

Marottoli, R.A. (2007). *Enhancement of Driving Performance Among Older Drivers*. Washington, DC: AAA Foundation for Traffic Safety.

Marottoli, R.A. (2008). Remediation from the physician's perspective. In D.W. Eby & L.J. Molnar (Eds.), *Proceedings of the North American License Policies Workshop*. Washington, DC: AAA Foundation for Traffic Safety.

Marottoli, R.A., Allore, H., Araujo, K.L., Iannone, L.P., Acampora, D., Gottschalk, M., Charpentier, P., Kasl, S., & Peduzzi, P. (2007). A randomized trial of a physical conditioning program to enhance the driving performance of older persons. *Journal of General Internal Medicine*, 22, 590–597.

Marottoli, R.A., Cooney, L.M., Wagner, D.R., Doucette, J., & Tinetti, M.E. (1994). Predictors of automobile crashes and moving violations among elderly drivers. *Annals of Internal Medicine*, 121, 842–846.

Marottoli, R.A. & Drickamer, M.A. (1993). Psychomotor mobility and the elderly driver. *Clinical Geriatric Medicine*, 9, 403–411.

Marottoli, R.A., Mendes de Leon, C.F., Glass, T.A., Williams, C.S., Cooney, L.M., Jr., & Berkman, L.F. (2000). Consequences of driving cessation: Decreased out-of-home activity levels. *Journal of Gerontology: Social Science*, 55B, S334–S340.

Marottoli, R.A., Mendes de Leon, C.F., Glass, T.A., Williams, C.S., Cooney, L.M. Jr., Berkman, L.F., & Tinetti, M.E. (1997). Driving cessation and increased depressive symptoms: Prospective evidence from the New Haven EPESE. Established populations for epidemiologic studies of the elderly. *Journal of the American Geriatrics Society*, 45, 202–206.

Marottoli, R.A., Ostfeld, A.M., Merrill, S.S., Perlman, G.D., Foley, D.J., & Cooney, L.M., Jr. (1993). Driving cessation and changes in mileage driven among elderly individuals. *Journal of Gerontology: Social Science*, 48, S255–S260.

Marottoli, R.A., Richardson, E.D., Stowe, M.H., Miller, E.G., Brass, L.M., Cooney, L.M., Jr., & Tinetti, M.E. (1998). Development of a test battery to identify older drivers at risk for self-reported adverse driving events. *American Geriatrics Society Journal*, 46, 562–568.

Marottoli, R.A., Van Ness, P.H., Araujo, K.L.B., Iannone, L.P., Acampora, D., Charpentier, P., & Peduzzi, P. (2007). A randomized trial of an education program to enhance older driver performance. *Journal of Gerontology: Medical Science*, 62A, 1113–1119.

Marshall, P.H., Elias, J.W., & Wright, J. (1985). Age related factors in motor error detection and correction. *Experimental Aging Research*, 11, 201–206.

Marshall, S.C. & Gilbert, N. (1999). Saskatchewan physicians' attitudes and knowledge regarding assessment of medical fitness to drive. *Canadian Medical Association Journal*, 160, 1701–1704.

Matlin, M.W. (1989). *Cognition, 2nd Edition*. Fort Worth, TX: Holt, Reinhart and Winston, Inc.

Matlin, M.W. & Foley, H.J. (1992). *Sensation and Perception*. Needham Heights, MA: Allyn and Bacon.

Mayhew, D.R. & Simpson, H.M. (2002). The safety value of driver education and training. *Traffic Injury Prevention*, 8, 3–8.

Mayhew, D.R., Simpson, H.M., & Ferguson, S.A. (2006). Collisions involving senior drivers: High-risk conditions and locations. *Traffic Injury Prevention*, 7, 117–124.

Maylor, E.A. (1991). Recognizing and naming tunes: Memory impairment in the elderly. *Journal of Gerontology: Psychological Sciences, 46*, 207–217.

Mayo Clinic. (2006). *Coronary Artery Disease: Risk Factors.* URL: http://www.mayoclinic. com/health/coronary-artery-disease/DS00064/DSECTION=4.

Mayo Clinic. (2007). *Epilepsy: Treatment.* URL: http://www.mayoclinic.com/health/epilepsy/ DS00342/DSECTION=8.

Mayo Clinic. (2008). *Coronary Failure: Causes.* URL: http://www.mayoclinic.com/health/ heart-failure/DS00061/DSECTION=3.

McCarthy, D. & Mann, W. (2006). Sensitivity and specificity of the assessment of driving-related skills older driver screening tool. *Topics in Geriatric Rehabilitation, 22*, 138–152.

McCloskey, L.W., Koepsell, T.D., Wolf, M.E., & Buchner, D.M. (1994). Motor vehicle collision injuries and sensory impairments of older drivers. *Age and Aging, 23*, 267–273.

McCoy, P.T., Tarawneh, M.S., Bishu, R.R., Ashman, R.D., & Foster, B.G. (1993). Evaluation of countermeasures for improving driving performance of older drivers. *Transportation Research Record*, 1405, 72–80.

McFarland, R.A. (1968). The sensory and perceptual processes in aging. In K.W. Schaie (Ed.), *Theory and Methods of Research in Aging.* Morgantown, WV: West Virginia University Press.

McFarland, R.A., Domey, R.G., Warren A.B., & Ward, D.C. (1960). Dark adaptation as a function of age: A statistical analysis. *Journal of Gerontology, 15*, 149–154.

McFarland, R.A. & Fischer, M.B. (1955). Alterations in dark adaptation as a function of age. *Journal of Gerontology, 10*, 424–428.

McGuckin, N.A. & Srinivasan, N. (2003). *Journey to Work in the United States: 1960–2000.* (Report No. FHWA-EP-03-058). Washington, DC: Federal Highway Administration.

McGwin, G., Jr. & Brown, D.B. (1999). Characteristics of traffic crashes among young, middle-aged, and older drivers. *Accident Analysis & Prevention, 31*, 181–198.

McGwin, G., Jr., Mayes, A., Joiner, W., DeCarlo, D.K., McNeal, S., & Owsley, C. (2004). Is glaucoma associated with motor vehicle collision involvement and driving avoidance? *Investigative Ophthalmology & Visual Science, 45*, 3934–3939.

McGwin, G., Jr., Sims, R.V., Pulley, L., & Roseman, J.M. (1999). Diabetes and automobile crashes in the elderly. *Diabetes Care, 22*, 220–227.

McGwin, G., Jr., Sims, R.V., Pulley, L., & Roseman, J.M. (2000). Relations among chronic medical conditions, medications, and automobile crashes in the elderly: A population-based case-control study. *American Journal of Epidemiology, 152*, 424–431.

McGwin, G., Jr., Xie, A., Mays, A., Joiner, W., DeCarlo, D.K., Hall, T.A., & Owsley, C. (2005). Visual field defects and risk of motor vehicle collisions among patients with glaucoma. *Investigative Ophthalmology & Visual Science, 46*, 4437–4441.

McKelvey, F.X. & Stamatiadis, N. (1988). Highway accident patterns in Michigan related to older drivers. *Transportation Research Record*, 1210, 53–57.

McKenzie, F. & Steen, T. (2002). A shocking impact. *Australian Planner, 39*, 194–201.

McKhann, G., Drachman, D., Folstein, M.F., Katzman, R., Price, D., & Stadlan, E.M. (1984). Clinical diagnosis of Alzheimer's disease: Report of the NINCDS-ADRDA work group. *Neurology, 34*, 939–944.

McKnight, A.J. (1988). Driver and pedestrian training. In *Transportation in an Aging Society: Volume 2. Improving Mobility and Safety for Older Persons.* Washington, DC: Transportation Research Board.

McKnight, A.J. & Adams, B.B. (1970). *Driver Education Task Analysis: Volume 1. Task Descriptions.* Alexandria, VA: Human Resource Research Organization.

McKnight, A.J. & McKnight, A.S. (1999). Multivariate analysis of age-related driver ability and performance deficits. *Accident Analysis & Prevention, 31*, 445–454.

McKnight, A.J., Shinar, D., & Hilburn, B. (1991). The visual and driving performance of monocular and binocular heavy-duty truck drivers. *Accident Analysis & Prevention*, 23, 225–237.

McPherson, K., Michael, J., Ostrow, A., & Shafron, P. (1988). *Physical Fitness and the Aging Driver: Phase I*. Washington, DC: AAA Foundation for Traffic Safety.

McRuer, D. & Weir, D.H. (1969). *Theory of Manual Vehicular Control*. Hawthorne, CA: Systems Technology, Inc.

Meindorfer, C., Körner, Y., Möller, J.C., Stiasny-Kolster, K., Oertel, W.H., & Krüger, H. (2005). Driving in Parkinson's disease: Mobility, accidents, and sudden onset of sleep at the wheel. *Movement Disorders*, 7, 832–842.

Mercier, C.R., Mercier, J.M., O'Boyle, M.W., & Strahan, R.F. (1997). *Validation of Relationship of Cognitive Skills Losses to Driving Performance*. Ames, IA: Midwest Transportation Center.

Meskin, J. (2002). *Knowledge Gaps and Needs Among Elderly Road Users in Drenthe: A Questionnaire Study* [in Dutch]. Leidschendam, Netherlands: SWOV.

Meuser, T.M., Carr, D.B., Berg-Weger, M., Niewoehner, P., & Morris, J.C. (2006). Driving and dementia in older adults: Implementation and evaluation of a continuing education project. *The Gerontologist*, 46, 680–687.

Meuser, T.M. (2008). License renewal policies and reporting of unfit drivers: Descriptive review and policy recommendations. In D.W. Eby & L.J. Molnar (Eds.), *Proceedings of the North American License Policies Workshop*. Washington, DC: AAA Foundation for Traffic Safety.

Michon, J.A. (1979). Dealing with danger: Report of the European Commission MRC workshop on physiology and psychological factors in performance under hazardous conditions (Report No. VK 79-01). Gieten, The Netherlands: Traffic Research Center, University of Groningen.

Michon, J.A. (1985). A critical view of driver behavior models: What do we know, what should we do? In *Human Behavior and Traffic Safety, Proceedings of a General Motors Symposium on Human Behavior and Traffic Safety*. New York, NY: Plenum Press.

Mihal, W.L. & Barrett, G.V. (1976). Individual differences in perceptual information processing and their relation to automobile accident involvement. *Journal of Applied Psychology*, 6, 229–233.

Miller, G.A. (1956). The magic number seven plus or minus two: Some limits on our capacity for processing information. *Psychology Review*, 63, 81–97.

Milton, K. (2004). Impact of public information on safe transportation of older people. In *Transportation in an Aging Society: A Decade of Experience*. Washington, DC: Transportation Research Board.

Mitchell, C.G.B. (1997). *The Potential of Intelligent Transportation Systems to Increase Accessibility to Transport for Elderly and Disabled People*. (Report No. TP 12926E). Montreal, Quebec: Transportation Development Centre.

Mitchell, R.K., Castleden, C.M., & Fanthome, Y. (1995). Driving, Alzheimer's disease and ageing: A potential cognitive screening device for all elderly drivers. *International Journal of Geriatric Psychiatry*, 10, 865–869.

Mohr, J.P. & Pessin, M.S. (1986). Extracranial carotid artery disease. In H.J.M. Barnett et al. (Eds.), *Stroke: Pathology, Diagnosis, and Management (Vol. II)*. New York, NY: Churchill Livingstone.

Mollenhauer, M.A., Dingus, T.A., & Hulse, M.C. (1995). *The Potential for Advanced Vehicle Systems to Increase the Mobility of Elderly Drivers*. Iowa City, IA: University of Iowa.

Molnar, F.J., Byszewski, A.M., Marshall, S.C., & Man-Son-Hing, M. (2005). In-office evaluation of medical fitness to drive: Practical approaches for assessing older people. *Canadian Family Physician*, 51, 372–379.

Molnar, L.J. & Eby, D.W. (2005). A brief look at driver license renewal policies in the United States. *Public Policy & Aging Report*, 15(2), 12–17.

Molnar, L.J. & Eby, D.W. (2008a). *Consensus-Based Recommendations From the North American License Policies Workshop.* Washington, DC: AAA Foundation for Traffic Safety.

Molnar, L.J. & Eby, D.W. (2008b). Getting around: Meeting the boomers' mobility needs. In R. Houston (Ed.), *Boomer or Bust? The New Political Economy of Aging.* Westport, CT: Praeger Publishing.

Molnar, L.J. & Eby, D.W. (2008c). The relationship between self-regulation and driving-related abilities in older drivers: An exploratory study. *Traffic Injury Prevention.*

Molnar, L.J., Eby, D.W., & Dobbs, B.M. (2005). Policy recommendations to the White House Conference on Aging Solutions Forum. *Public Policy & Aging Report,* 15(2), 24–27.

Molnar, L.J., Eby, D.W., & Miller, L.L. (2003). *Promising Approaches for Enhancing Elderly Mobility.* Ann Arbor, MI: University of Michigan Transportation Research Institute.

Molnar, L.J., Eby, D.W., St. Louis, R.M., & Neumeyer, A.L. (2007). *Promising Approaches for Promoting Lifelong Community Mobility.* Washington, DC: AARP.

Monaco, K. & Pellerito, J.M., Jr. (2006). Funding for driver rehabilitation services and equipment. In J.M. Pellerito (Ed.), *Driver Rehabilitation and Community Mobility.* St. Louis, MO: Elsevier Mosby.

Moore, L.W. & Miller, M. (2005). Driving strategies used by older adults with macular degeneration: Assessing the risks. *Applied Nursing Research,* 18, 110–116.

Morrisey, M.A. & Grabowski, D.C. (2005). State motor vehicle laws and older drivers. *Health Economics,* 14, 407–419.

Moskowitz, H. & Smiley, A. (1982). Effects of chronically-administered buspirone and diazepam on driving-related skills performance. *Journal of Clinical Psychiatry,* 43, 45–55.

Murray-Leslie, C. (1991). Driving for the person disabled by arthritis. *Journal of Rheumatology,* 30, 54–55.

Myers, R.S., Ball, K.K., Kalina, T.D., Roth, D.L., & Goode, K.T. (2000). The relationship of useful field of view and other screening instruments to on-road driving performance. *Perceptual and Motor Skills,* 91, 279–290.

Nahin, R.L., Fitzpatrick, A.L., Williamson, J.D., Burke, G.L., DeKosky, S.T., & Furberg, C. (2006). Use of herbal medicine and other dietary supplements in community-dwelling older people: Baseline data from the ginko evaluation of memory study. *Journal of the American Geriatrics Society,* 54, 1725–1735.

Nakra, B.R.S., Gfeller, J.D., & Hassan, R. (1992). A double-blind comparison of the effects of temazepam and triazolam on residual, daytime performance in elderly insomniacs. *International Psychogeriatrics,* 4, 45–53.

Nasvadi, G.E. & Vavrik, J. (2007). Crash risk of older drivers after attending a mature driver education program. *Accident Analysis & Prevention,* 39, 1073–1079.

National Eye Institute. (2008a). *Facts About Cataract.* URL: http://www.nei.nih.gov/health/cataract/cataract_facts.asp.

National Eye Institute. (2008b). *Facts About Retinopathy.* URL: http://www.nei.nih.gov/health/diabetic/retinopathy.asp.

National Eye Institute. (2008c). *Facts About Glaucoma.* URL: http://www.nei.nih.gov/health/glaucoma/glaucoma_facts.asp.

National Eye Institute. (2008d). *Facts About Macular Degeneration.* URL: http://www.nei.nih.gov/health/maculardegen/armd_facts.asp.

National Highway Traffic Safety Administration. (1993). *Addressing the Safety Issues Related to Younger and Older Drivers: A Report to Congress* (Report No. DOT HS 807-957). Washington, DC: U.S. Department of Transportation.

National Highway Traffic Safety Administration. (2003). *Safe Driving for Older Adults.* (Report No. DOT HS 809 493). Washington, DC: U.S. Department of Transportation.

National Highway Traffic Safety Administration. (2007). *Adapting Motor Vehicles for Older Drivers.* (Report No. DOT HS 810 732). Washington, DC: U.S. Department of Transportation.

National Highway Traffic Safety Administration. (2008). *Traffic Safety Facts, 2006 Data: Pedestrians*. (Report No. DOT HS 810 810). Washington, DC: U.S. Department of Transportation.

National Institute of Neurological Disorders and Stroke. (1990). Special Report: Classification of cerebrovascular disease. *Stroke*, 21, 637–676.

National Institute of Neurological Disorders and Stroke. (2004). *Parkinson's Disease Backgrounder*. URL: http://www.ninds.nih.gov/disorders/parkinsons_disease/parkinsons_disease_backgrounder.htm.

National Institutes of Health. (1996). *Congestive Heart Failure Data Fact Sheet*. URL: http://www.wrongdiagnosis.com/artic/nhlbi_congestive_heart_failure_data_fact_sheet_nhlbi.htm.

National Institutes of Health (1999). *Data Fact Sheet: Asthma Statistics*. URL: http://www.nhlbi.nih.gov/health/prof/lung/asthma/asthstat.pdf

National Institutes of Health. (2008). *What Is Asthma?* URL: http://www.nhlbi.nih.gov/health/dci/Diseases/Asthma/Asthma_WhatIs.html.

National Multiple Sclerosis Society. (2005). Epidemiology. In *National MS Society Information Source Book*. New York, NY: National MS Society.

National Multiple Sclerosis Society. (2006). *About MS: What Is Multiple Sclerosis?* URL: http://main.nationalmssociety.org.

National Parkinson's Foundation. (2008). *About Parkinson's Disease*. URL: http://www.parkinson.org/.

National Study to Prevent Blindness. (1980). *Vision Problems in the U.S. Data Analysis: Definitions, Data Sources, Detailed Data Tables, Analysis, Integration*. New York, NY: National Society to Prevent Blindness.

Nebes, R.D. & Madden, D.J. (1988). Different patterns of cognitive slowing produced by Alzheimer's disease and normal aging. *Psychology & Aging*, 3, 102–104.

Neutel, C.I. (1995). Risk of traffic accident injury after a prescription for a benzodiazepine. *Annals of Epidemiology*, 5, 239–244.

New York State Office of the Aging. (2000). *When You Are Concerned: A Guide for Family, Friends, and Caregivers Concerned About the Safety of an Older Driver*. New York, NY: New York State Office of the Aging.

Nicholson, A.N. (1986). Hypnotics and transient insomnia. In J.F. O'Hanlon & J.J. De Gier (Eds.), *Drugs and Driving*. Basingstoke, Hampshire: Taylor & Francis Ltd.

Norman, D.A. & Bobrow, D.G. (1975). On data-limited and resource-limited processes. *Cognitive Psychology*, 7, 44–65.

Norman, J.F., Dawson, T.E., & Butler, A.K. (2000). The effects of age upon the perception of depth and 3-D shape from differential motion and binocular disparity. *Perception*, 29, 1335–1359.

Norman, J.F., Ross, H.E., Hawkes, L.M., & Long, J.R. (2003). Aging and the speed of perception. *Perception*, 32, 85–96.

O'Brien, A.P. & Richardson, E. (1985). Use of roundabouts in Australia. *55th Annual Compendium of Technical Papers*. Washington, DC: Institute of Transportation Engineers.

Odenheimer, G.L., Beaudet, M., Gette, A.M., Albert, M.S., Grande, L., & Minaker, K.L. (1994). Performance-based driving evaluation of the elderly driver: Safety, reliability and validity. *Journal of Gerontology: Medical Science*, 49, 153–159.

Ogden, J.A. (1990). Spatial abilities and deficits in aging and age-related disorders. In F. Boller & J. Grafman (Eds.), *Handbook of Neuropsychology*. New York, NY: Elsevier.

Ogden, M.A., Womak, K.N., & Mounce, J.M. (1990). Motorist comprehension of signing in urban arterial work zones. *Transportation Research Record*, 1281, 127–135.

O'Hanlon, J.F. (1992). Discussion of 'Medications and the safety of the older driver' by Ray et al. *Human Factors*, 34, 49–51.

O'Hanlon, J.F. & de Gier, J.J. (Eds.). (1986). *Drugs and Driving*. Basingstoke, UK: Taylor & Francis Ltd.

O'Hanlon, J.F., Haak, T.W., Blaauw, G.J., & Riemersma, J.B.J. (1982). Diazepam impairs lateral position control in highway driving. *Science, 217*, 79–81.

Ohme, P.J. & Schnell, T. (2001). Is wider better? Enhancing pavement marking visibility for older drivers. In *Proceedings of the Human Factors and Ergonomics Society Annual Meeting*. Santa Monica, CA: Human Factors and Ergonomics Society.

Ohta, R.J. (1983). Spatial orientation in the elderly: The current status of understanding. In H.L. Pick & L.P. Acredelo (Eds.), *Spatial Orientation: Theory, Research, and Application*. New York, NY: Plenum Press.

Olson, P.L., Sivak, M., & Egan, J.C. (1983). *Variables Influencing the Nighttime Legibility of Highway Signs*. (Report No. UMTRI-83-36). Ann Arbor, MI: University of Michigan Transportation Research Institute.

Olzak, L.A. & Thomas, J.P. (1986). Seeing spatial patterns. In K.R. Boff, L. Kaufman, & J.P. Thomas (Eds.), *Handbook of Perception and Human Performance: Volume I. Sensory Processes and Perception*. New York, NY: John Wiley and Sons.

Ong, K.L., Cheung, B.M.Y., Man, Y.B., Lau, C.P., & Lam, S.L. (2007). Prevalence, awareness, treatment, and control of hypertension among United States adults 1999–2004. *Hypertension, 49*, 69–75.

Ostrow, A.C., Shaffron, P., & McPherson, K. (1992). The effects of a joint range-of-motion physical fitness training program on the automobile driving skills of older adults. *Journal of Safety Research, 23*, 207–219.

Ott, R., Heindel, W.C., Whelihan, W.M., Caron, M.D., Piatt, A.L., & DiCarlo, M.A. (2003). Maze test performance and reported driving ability in early Dementia. *Journal of Geriatric Psychiatry and Neurology, 16*, 151–155.

Owsley, C. (2004). Impact of an educational program on the safety of high-risk, visually impaired older drivers. *American Journal of Preventive Medicine, 26*, 222–229.

Owsley, C. & Ball, K. (1993). Assessing visual function in the older driver. *Clinical Geriatric Medicine, 9*, 389–401.

Owsley, C., Ball, K., McGwin, G., Jr., Sloane, M.E., Roenker, D.L., White, M.F., & Overley, T.E. (1998). Visual processing impairment and risk of motor vehicle crash among older adults. *Journal of the American Medical Association, 279*, 1083–1088.

Owsley, C., Ball, K., Sloane, M.E., Roenker, D.L., & Bruni, J.R. (1991). Visual/cognitive correlates of vehicle accidents in older drivers. *Psychology & Aging, 6*, 403–415.

Owsley, C. & McGwin, G., Jr. (1999). Vision impairment and driving. *Survey of Ophthalmology, 43*, 535–550.

Owsley, C., McGwin, G., Jr., & Ball, K. (1998). Vision impairment, eye disease, and injurious motor vehicle crashes in the elderly. *Ophthalmic Epidemiology, 5*, 101–113.

Owsley, C., McGwin, G., Mays, A., Joiner, W., DeCarlo, D.K., & McNeal, S. (2004). Is glaucoma associated with motor vehicle collision involvement and driving avoidance? *Investigative Ophthalmology & Visual Science, 45*, 1123.

Owsley, C., McGwin, G., Jr., Phillips, J.M., McNeal, S.F., & Stalvey, B.T. (2004). Impact of an educational program on the safety of high-risk, visually impaired, older drivers. *American Journal of Preventive Medicine, 26*, 222–229.

Owsley, C., McGwin, G., Jr., Sloane, M., Wells, J., Stalvey, B.T., & Gauthreaux, S. (2002). Impact of cataract surgery on motor vehicle crash involvement by older adults. *The Journal of the American Medical Association, 288*, 841–849.

Owsley, C., Sekuler, R., & Siemsen, D. (1983). Contrast sensitivity throughout adulthood. *Vision Research, 23*, 689–699.

Owsley, C. & Sloane, M.E. (1990). Vision and aging. In F. Bolter & J. Grafman (Eds.), *Handbook of Neuropsychology: Volume 4*. Amsterdam, The Netherlands: Elsevier.

Owsley, C., Stalvey, B.T., & Phillips, J.M. (2003). The efficacy of an educational intervention in promoting self-regulation among high-risk older drivers. *Accident Analysis &
Prevention*, 35, 393–400.

Owsley, C., Stalvey, B., Wells, J., & Sloane, M.E. (1999). Older drivers and cataract: Driving
habits and crash risk. *Journals of Gerontology Series A—Biological Sciences & Medical
Sciences*, 54, 203–211.

Owsley, C., Stalvey, B., Wells, J., Sloane, M.E., & McGwin, G., Jr. (2001). Visual risk factors
for crash involvement in older drivers with cataract. *Archives of Ophthalmology*, 119,
881–887.

Oxley, J., Charlton, J., & Fildes, B. (2005). *The Effect of Cognitive Impairment on Older
Pedestrian Behaviour and Crash Risk*. Victoria, Australia: Monash University Accident
Research Centre.

Oxley, J., Fildes, B., Corben, B., & Langford, J. (2006). Intersection design for older drivers.
Transportation Research Part F, 9, 335–346.

Oxley, P.R., Barham, P.A., & Ayala, B.E. (1995). The use of route guidance systems by elderly
drivers. In *Proceedings of the 1994 World Congress on Applications of Transport
Telemetrics and Intelligent Vehicle-Highway Systems*. Boston, MA: Artech House.

Parasuraman, R. (1986). Vigilance, monitoring, and search. In K.R. Boff, L. Kaufman, & J.P.
Thomas (Eds.), *Handbook of Perception and Human Performance: Volume 2. Cognitive
Processes and Performance*. New York, NY: John Wiley and Sons.

Parasuraman, R. (1991). Attention and driving performance in Alzheimer's dementia.
*Proceedings of the Conference: Strategic Highway Research Program and Traffic Safety
on Two Continents*. Gothenburg, Sweden.

Parkinson, S.R., Lindholm, J.M., & Inman, V.W. (1982). An analysis of age differences in
immediate recall. *Journal of Gerontology*, 37, 425–431.

Parrish, R.K., Gedde, S.J., Scott, I.U., Feuer, W.J., Schiffman, J.C., Mangione, C.M., &
Montenegro-Piniella, A. (1997). Visual function and quality of life among participants
with glaucoma. *Archives of Ophthalmology*, 115, 1447–1455.

Patlak, M. (1990). Light for sight; lasers beginning to solve vision problems. *FDA Consumer*,
July/August, 15–18.

Pennsylvania Department of Transportation. (2006). *Talking With Older Drivers: A Guide for
Family and Friends*. (Report No. 345-1-06). Harrisburg, PA: Pennsylvania Department
of Transportation.

Perel, M. (1998). Helping older drivers benefit from in-vehicle technologies. In *Proceedings
of the 16th International Technical Conference on the Enhanced Safety of Vehicles*.
Washington DC: U.S. Department of Transportation.

Perlmutter, M. & Nyquist, L. (1990). Relationships between self-reported physical and mental health and intelligence performance across adulthood. *Journal of Gerontology:
Psychological Sciences*, 45, 145–155.

Persaud, B.N., Retting, R.A., Garder, P.E., & Lord, D. (2000). *Crash Reduction Following
Installation of Roundabouts in the United States*. Arlington, VA: Insurance Institute for
Highway Safety.

Persson, D. (1993). The elderly driver: Deciding when to stop. *The Gerontologist*, 33, 88–91.

Peterson, L.R. & Peterson, M.J. (1959). Short-term retention of individual verbal items.
Journal of Experimental Psychology, 58, 495–512.

Petrofsky, J.S. & Lind, A.R. (1975). Aging, isometric strength and endurance, and cardiovascular responses to static effort. *Journal of Applied Psychology*, 1, 91–95.

Petzäll, J. (1995). The design of entrances of taxis for elderly and disabled passengers. *Applied
Ergonomics*, 26, 343–352.

Pietrucha, M.T. (1995). Human factors issues related to work zone safety. *Transportation
Builder*, 7, 409–442.

Pigman, J.G. & Agent, K.R. (1990). Highway accidents in construction and maintenance work zones. *Transportation Research Record*, 1270, 12–21.

Pike, J. (2004). Reducing injuries and fatalities to older drivers: Vehicle concepts. In *Transportation in an Aging Society: A Decade of Experience*. Washington, DC: Transportation Research Board.

Polzin, S. E. & Chu, X. A. (2005). A closer look at public transportation mode share trends. *Journal of Transportation & Statistics*, 8, 41–53.

Ponds, R.W.H.M., Brouwer, W.H., & van Wolffelaar, P.C. (1988). Age differences in divided attention in a simulated driving task. *Journal of Gerontology*, 43, 151–156.

Popkin, C.L. & Waller, P.F. (1989). Epilepsy and driving in North Carolina: An exploratory study. *Accident Analysis & Prevention*, 21, 389–393.

Porter, M.M. & Whitton, M.J. (2002). Assessment of driving with global positioning system and video technology in young, middle-aged, and older drivers. *Journal of Gerontology*, 57A, M578–M582.

Posse, C., McCarthy, D. P., & Mann, W.C. (2006). A pilot study of interrater reliability of the Assessment of Driving-Related Skills: Older driver screening tool. *Topics in Geriatric Rehabilitation*, 22, 113–120.

Potts, I., Stutts, J., Pfefer, R., Neuman, T.R., Slack, K.L., & Hardy, K.K. (2004). *Guidance for Implementation of the AASHTO Strategic Highway Safety Plan—Volume 9: A Guide for Reducing Collisions Involving Older Drivers*. (NCHRP Report 500). Washington, DC: Transportation Research Board.

Preusser, D.F., Williams, A.F., Ferguson, S.A., Ulmer, R.G., & Weinstein, H.B. (1998). Fatal crash risk for older drivers at intersections. *Accident Analysis & Prevention*, 30, 151–159.

Prevent Blindness America. (2005). *Color Blindness*. URL: http://www.preventblindness.org/eye_problems/colorvision.html.

Pucher, J. & Dijkstra, L. (2003). Promoting safe walking and cycling to improve public health: Lessons from the Netherlands and Germany. *American Journal of Public Health*, 93(9), 1509–1516.

Rabbitt, P. (1965). An age-decrement in the ability to ignore irrelevant information. *Journal of Gerontology*, 20, 233–238.

Rabbitt, P. (1989). Inner-city decay? Age changes in structure and process in recall of familiar topographical information. In L.W. Poon, D.C. Rubin, & B.A. Wilson (Eds.), *Everyday Cognition in Adulthood and Late Life*. Cambridge, MA: Cambridge University Press.

Rabbitt, P., Carmichael, A., Jones, S., & Holland, C.A. (1996). *When and Why Older Drivers Give Up Driving*. Manchester, UK: AAA Foundation for Road Safety Research.

Racette, L. & Casson, E.J. (2005). The impact of visual field loss on driving performance: Evidence from on-road driving assessments. *Optometry and Vision Science*, 82, 668–674.

Radford, K.A., Lincoln, N.B., & Lennox, G. (2004). The effects of cognitive abilities on driving in people with Parkinson's disease. *Disability and Rehabilitation*, 26, 65–70.

Ragland, D., Satariano, W.A., & MacLeod, K. E. (2004). Reasons given by older people for limitation or avoidance of driving. *The Gerontologist*, 44, 237–244.

Ragland, D.R., Satariano, W.A., & MacLeod, K.E. (2005). Driving cessation and depressive symptoms. *Journal of Gerontology: Medical Sciences*, 60A, 399–403.

Raitanen, T., Tormakangas, T., Mollenkopf, H., & Marcellini, F. (2003). Why do older drivers reduce driving? Findings from three European countries. *Transportation Research Part F*, 6, 81–95.

Ram, N., Talmor, D., & Brasel, K. (2005). Predicting significant torso trauma. *The Journal of Trauma*, 59, 132–135.

Ramaekers, J.G. (2003). Antidepressants and driver impairment: Empirical evidence from a standard on-the-road test. *Journal of Clinical Psychiatry*, 64, 20–29.

Ranney, T.A. (1994). Models of driving behavior: A review of their evolution. *Accident Analysis & Prevention, 26, 733–750.*

Ranney, T.A., Garrott, R., & Goodman, M.J. (2001). NHTSA driver distraction research: Past, present, and future. In *Proceedings of the 17th International Technical Conference on the Enhanced Safety of Vehicles.* (Report No. 233, CD-ROM). Washington, DC: U.S. Department of Transportation.

Ranney, T.A. & Pulling, N.H. (1989). Relation of individual differences in information-processing ability to driving performance. *Proceedings of the Human Factors Society, 33rd Annual Meeting.* Santa Monica, CA: Human Factors Society.

Ranney, T.A. & Pulling, N.H. (1990). Performance differences on driving and laboratory tasks between drivers of different ages. *Transportation Research Record, 1281,* 3–10.

Rasmussen, J. (1980). The human as a system's component. In H.T. Smith & T.R. Greed (Eds.), *Human Interaction With Computers.* London, UK: Academic Press.

Rasmussen, J. (1987). The definition of human error and a taxonomy for technical system design. In J. Rasmussen, K. Duncan, J. Leplat (Eds.), *New Technology and Human Error.* Chichester, UK: Wiley.

Ray, W.A., Fought, R.L., & Decker, M.D. (1992). Psychoactive drugs and the risk of injurious motor vehicle crashes in elderly drivers. *American Journal of Epidemiology, 136,* 873–883.

Ray, W.A., Gurwitz, J.H., Decker, M.D., & Kennedy, D.L. (1992). Medications and the safety of the older driver: Is there a basis for concern? *Human Factors, 34,* 33–47.

Ray, W.A., Thapa, P.B., & Shorr, R.I. (1993). Medications and the older driver. In S. Retchin (Ed.), *Medical Considerations of the Older Driver, Clinics in Geriatric Medicine, 9,* 413–438.

Raytheon Commercial Infrared and EICAN-Texas Optical Technology. (2000). *NightDriver™ Thermal Imaging Camera and HUD Development Program for Collision Avoidance Applications* (Report DOT-HS-809-163). Washington, DC: U.S. Department of Transportation.

Reason, J., Manstead, A.S.R., Stradling, S.G., Parker, D., & Baxter, J. (1991). *The Social and Cognitive Determinants of Aberrant Driving Behavior.* Crowthorne, UK: UK Transport Research Laboratory.

Redepenning, S. (2006). *Driver Rehabilitation Across Age and Disability: An Occupational Therapy Guide.* Bethesda, MD: American Occupational Therapy Association.

Regestein, Q.R. (1992). Treatment of insomnia in the elderly. In C. Salzman (Ed.), *Clinical Geriatric Psychopharmacology.* New York, NY: McGraw-Hill.

Rehm, C.G. & Ross, S.E. (1995). Syncope as etiology of road crashes involving elderly drivers. *American Surgeon, 61,* 1006–1008.

Retting, R.A., Kyrychenko, S.Y., & McCartt, A.T. (2007). Long-term trends in public opinion following construction of roundabouts. *Transportation Research Record, 2019,* 219–224.

Richardson, E.D. & Marottoli, R.A. (2003). Visual attention and driving behaviors among community-living older persons. *Journals of Gerontology, 58A,* 832–836.

Rimini-Doering, M., Altmueller, T., Ladstaetter, U., & Rossmeier, M. (2005). Effects of lane departure warning on drowsy drivers' performance and state in a simulator. In *Proceedings of the Third International Driving Symposium on Human Factors in Driver Assessment, Training, and Vehicle Design.* Ames, IA: The University of Iowa.

Rinalducci, E.J., Mouloua, M., & Smither, J. (2001). *Cognitive and Perceptual Factors in Aging and Driving Performance.* (Report No. VPL-03-01). Orlando, FL: University of Central Florida, Visual Performance Laboratory.

Ritter, G. & Steinberg, H. (1979). Parkinsonismus und Fahrtauglichkeit [in German]. *MMW Much Med Wochenschr, 121,* 1329–1330.

Rizzo, M., McGehee, D.V., Dawson, J.D., & Anderson, S.N. (2001). Simulated car crashes at intersections in drivers with Alzheimer disease. *Alzheimer Disease and Associated Disorders*, 15, 10–20.

Rizzo, M., Reinach, S., McGehee, D., & Dawson, J. (1997). Simulated car crashes and crash predictors in drivers with Alzheimer disease. *Archives of Neurology*, 54, 545–551.

Roenker, D.L., Cissell, G.M., Ball, K.K., Wadley, V.G., & Edwards, J.D. (2003). Speed-of processing and driving simulator training result in improved driving performance. *Human Factors*, 45, 218–233.

Rosenbloom, S. (2003). *The Mobility Needs of Older Americans: Implications for Transportation Reauthorization*. Washington, DC: Brookings Institution Center on Urban and Metropolitan Policy.

Rosenbloom, S. (2004). Mobility of the elderly: Good news and bad news. In *Transportation in an Aging Society: A Decade of Experience*. Washington, DC: Transportation Research Board.

Rothe, J.P. (1990). *The Safety of Elderly Drivers: Yesterday's Young in Today's Traffic*. New Brunswick, NJ: Transaction Publishers.

Rubin, G.S., Hg, E.S.W., Bandeen-Roche, K., Keyl, P.M., Freeman, E.E., West, S.K., & SEE Project Team (2007). A prospective, population-based study of the role of visual impairment in motor vehicle crashes among older drivers: The SEE Study. *Investigative Ophthalmology & Visual Science*, 48, 1483–1491.

Rubinsztein, J. & Lawton, C.A. (1995). Depression and driving in the elderly. *International Journal of Geriatric Psychiatry*, 10, 15–17.

Ruch, P.L. (1934). The differentiative effects of age upon human learning. *Journal of General Psychology*, 11, 261–286.

Rudin-Brown, C. & Parker, H.A. (2004). Behavioural adaptation to adaptive cruise control (ACC): Implications for preventive strategies. *Transportation Research Part F*, 7, 59–76.

Ruechel, S. & Mann, W.C. (2005). Self-regulation of driving by older persons. *Physical & Occupational Therapy in Geriatrics*, 23, 91–101.

Rumar, K. (2002). *Night Vision Enhancement Systems: What Should They Do and What More Do We Need to Know?* (Report No. UMTRI-2002-12). Ann Arbor, MI: University of Michigan Transportation Research Institute.

Ruppel, R., Schluter, C.A., Boczor, S., Meinertz, T., Schluter, M., Kuck, K.H., & Cappato, R. (1998). Ventricular tachycardia during follow-up in patients resuscitated from ventricular fibrillation: Experience from stored electrograms of implantable cardioverter-defibrillators. *Journal of the American College of Cardiology*, 32, 1724–1730.

Ryan, E.B. (1992). Beliefs about memory changes across the adult lifespan. *Journal of Gerontology: Psychological Sciences*, 47, 41–46.

Ryan, G.A., Legge, M., & Rosman, D. (1998). Age related crashes in drivers' crash risk and crash type. *Accident Analysis & Prevention*, 30, 379–387.

Said, F.S. & Weale, R.A. (1959). The variation with age of the spectral transmissivity of the living human crystalline lens. *Gerontologia*, 3, 213–231.

Salthouse, T.A. (1980). Age and memory: Strategies for localizing the loss. In L.W. Poon, J.L. Fozard, L.S. Cermak, D. Arenberg, & L.W. Thompson (Eds.), *New Directions in Memory and Aging: Proceedings of the George A. Talland Memorial Conference*. Hillsdale, NJ: Lawrence Erlbaum Associates.

Salthouse, T.A. (1987). Adult age differences in integrative spatial ability. *Psychology & Aging*, 2, 254–260.

Salthouse, T.A. (1990). Working memory as a processing resource in cognitive aging. *Developmental Review*, 10, 101–124.

Salthouse, T.A., Mitchell, D.R., Skovronek, E., & Babcook, R.L. (1989). Effects of adult age and working memory on reasoning and spatial ability. *Journal of Experimental Psychology: Learning, Memory, & Cognition*, 15, 507–516.

Salthouse, T.A. & Skovronek, E. (1992). Within-context assessment of age differences in working memory. *Journal of Gerontology: Psychological Sciences,* 47, 110–120.

Salzman, C. (1992). Treatment of anxiety. In C. Salzman (Ed.), *Clinical Geriatric Psychopharmacology.* New York, NY: McGraw-Hill.

Sanders, A.F. (1986). Drugs, driving and the measurement of human performance. In J.F. O'Hanlon & J.J. Gier (Eds.), *Drugs and Driving.* Basingstoke, UK: Taylor & Francis Ltd.

Sarks, S.H. (1975). The aging eye. *Medical Journal of Australia,* 2, 602–604.

Savage, D.D., Corwin, L., McGee, D.L., Kannel, W.B., & Wolf, P.A. (1985). Epidemiological features of isolated syncope: The Framingham study. *Stroke,* 16, 626–629.

Schanke, A. & Sundet, K. (2000). Comprehensive driving assessment: Neuropsychological testing and on-road evaluation of brain injured patients. *Scandinavian Journal of Psychology,* 41, 113–121.

Schanke, A.K., Grismo, J., & Sundet, K. (1995). Multiple sclerosis and prerequisites for driver's license: A retrospective study of 33 patients with multiple sclerosis assessed at Sunnaas hospital [in Norwegian]. *Tidsskrift for Den Norske Laegeforening,* 115, 1349–1352.

Schieber, F. (1994). *Recent Developments in Vision, Aging, and Driving: 1988–1994* (Report No. UMTRI-94-26). Ann Arbor, MI: The University of Michigan Transportation Research Institute.

Schieber, F., Hiris, E., White, J., Williams, M., & Brannan, J. (1990). Assessing age-differences in motion perception using simple oscillatory displacement versus random dot cinematography. *Investigative Ophthalmology & Visual Science,* S31, 355.

Schieber, F., Kline, D.W., Kline, T.J.B., & Fozard, J.L. (1992). *The Relationship Between Contrast Sensitivity and the Visual Problems of Older Drivers* (SAE Technical Paper No. 920613). Warrendale, PA: Society of Automotive Engineers, Inc.

Schiff, W., Oldak, R., & Shah, V. (1992). Aging persons' estimates of vehicular motion. *Psychology & Aging,* 7, 518–525.

Schmidt, P., Haarhoff, K., & Bonte, W. (1990). Sudden natural death at the wheel—A particular problem of the elderly? *Forensic Science International,* 48, 155–162.

Schold-Davis, E. (2008). Remediation from the occupational therapist's perspective. In D.W. Eby & L.J. Molnar (Eds.), *Proceedings of the North American License Policies Workshop.* Washington, DC: AAA Foundation for Traffic Safety.

Schonfield, A.E.D. (1969). *In Search of Early Memories.* Paper presented at the International Congress of Gerontology, Washington, DC.

Schultheis, M.T., Garay, E., & DeLuca, J. (2001). The influence of cognitive impairment on driving performance in multiple sclerosis. *Neurology,* 56, 1089–1094.

Schultheis, M.T., Garay, E., Millis, S.R., & DeLuca, J. (2002). Motor vehicle crashes and violation among drivers with multiple sclerosis. *Archives of Physical Medical Rehabilitation,* 83, 1175–1178.

Schulze, H. (1990). *Lifestyle, Leisure Style and Traffic Behaviours of Young Drivers.* (Report No. VTI-364). Linköping, Sweden: Swedish Road and Traffic Research Institute.

Scialfa, C.T., Guzy, L.T., Leibowitz, H.W., Garvey, P.M., & Tyrrell, R.A. (1991). Age differences in estimating vehicle velocity. *Psychology & Aging,* 6, 60–66.

Scialfa, C.T., Kline, D.W., & Lyman, B.J. (1987). Age differences in target identification as a function of retinal location and noise level: Examination of the useful field of view. *Psychology & Aging,* 2, 14–19.

Scilley, K., Jackson, G.R., Cideciyan, A.V., Maguire, M.G., Jacobson, S.G., & Owsley, C. (2002). Early age-related maculopathy and self-reported visual difficulty in daily life. *Ophthalmology,* 109, 1235–1242.

Sekuler, R. & Ball, K. (1986). Visual localization: Age and practice. *Journal of the Optical Society of America,* 3, 864–867.

Sekuler, R. & Blake, R. (1985). *Perception.* New York, NY: Alfred A. Knopf, Inc.

Seppala, T., Linnoila, M., Elonen, E., Mattila, M.J., & Maki, M. (1975). Effect of tricyclic anti-depressants and alcohol on psychomotor skills related to driving. *Clinical Pharmacology and Therapeutics*, 17, 515–522.

Shaheen, S.A. & Neimeier, D.A. (2001). Integrating vehicle design and human factors: Minimizing elderly driving constraints. *Transportation Research Part C*, 9, 155–174.

Shankar, V.N., Sittikariya, S., & Shyu, M. (2006). Some insights on roadway infrastructure design for safe elderly pedestrian travel. *International Association of Traffic Safety and Science Research*, 30, 21–26.

Sharpe, J.A. & Sylvester, T.O. (1978). Effects of age on horizontal smooth pursuit. *Investigative Ophthalmology & Visual Science*, 17, 465–468.

Shawaryn, M.A., Schulthesis, M.T., Garay, E., & DeLuca, J. (2002). Assessing functional status: Exploring the relationship between the Multiple Sclerosis Functional Composite and driving. *Archives of Physical Medical Rehabilitation*, 83, 1123–1129.

Shechtman, O., Classen, S., Stephens, B., Bendixen, R., Belchoir, P., Sandhu, M., McCarthy, D., Mann, W., & Davis, E. (2007). The impact of intersection design on simulated driving performance of young and senior adults. *Traffic Injury Prevention*, 8, 78–86.

Sheldon, R. & Koshman, M.L. (1995). Can patients with neuromediated syncope safely drive motor vehicles? *American Journal of Cardiology*, 75, 955–956.

Shepard, R.J. (1998). Aging and exercise. In T.D. Fahey (Ed.), *Encyclopedia of Sport Medicine and Science*. URL: http:\\sportsci.org

Shepard, R.N. & Metzler, J. (1971). Mental rotation of three-dimensional objects. *Science*, 171, 701–703.

Shinar, D. (1977). *Driver Visual Limitation Diagnosis and Treatment* (Report No. DOT-HS-803-260). Washington, DC: U.S. Department of Transportation.

Shinar, D. (1978). *Psychology on the Road: The Human Factor in Traffic Safety*. New York, NY: Wiley.

Shinar, D., Dewar, R.E., Summala, H., & Zakowska, L. (2003). Traffic sign symbol comprehension: A cross-cultural study. *Ergonomics*, 46, 1549–1565.

Siegler, R.S. (1991). *Children's Thinking, 2nd Edition*. Englewood Cliffs, NJ: Prentice Hall.

Silverstein, N.M. (2008). Alzheimer's disease and fitness to drive. In D.W. Eby & L.J. Molnar (Eds.), *Proceedings of the North American License Policies Workshop*. Washington, DC: AAA Foundation for Traffic Safety.

Silverstein, N.M., Flaherty, G., & Tobin, T.S. (2002). *Dementia and Wandering Behavior: Concern for the Lost Elder*. New York, NY: Springer.

Silverstein, N.M., Gottlieb, A.S., & Van Ranst, E. (2005). Use of a video intervention to increase elders' awareness of low cost vehicle modifications to enhance driving safety and comfort. *Transportation Research Record*, 1922, 15–20.

Silverstone, T. (1998). The influence of psychiatric disease and its treatment on driving performance. *International Clinical Psychopharmacology*, 3, 59–66.

Sims, R.V., McGwin, G., Jr., Allman, R.M., Ball, K., & Owsley, C. (2000). Exploratory study of incident vehicle crashes among older drivers. *Journal of Gerontology*, 55A, M22–M27.

Sims, R.V., Owsley, C., Allman, R.M., Ball, K., & Smoot, T.M. (1998). A preliminary assessment of the medical and functional factors associated with vehicle crashes by older adults. *Journal of the American Geriatrics Society*, 46, 556–561.

Singh, R., Pentland, B., Hunter, J., & Provan, F. (2006). Parkinson's disease and driving ability. *Journal of Neurology, Neurosurgery, and Psychiatry*, 78, 363–366.

Siren, A. & Hakamies-Blomqvist, L. (2005). Sense and sensibility. A narrative study of older women's car driving. *Transportation Research Part F*, 8, 213–228.

Siren, A., Hakamies-Blomqvist, L., & Lindeman, M. (2004). Driving cessation and health in older women. *Journal of Applied Gerontology*, 23, 58–69.

Sisk, J.E. (2006). Antihistamines. *Gale Encyclopedia of Children's Health*. Detroit, MI: Thomson Gale.

Sivak, M. (1996). The information drivers' use: Is it indeed 90 percent visual? *Perception*, 25, 1081–1089.

Sivak, M., Campbell, K.L., Schneider, L.W., Sprague, J.K., Streff, F.M., & Waller, P.F. (1995). The safety and mobility of older drivers: What we know and promising research issues. *UMTRI Research Review*, 26(1), 1–21.

Sivak, M., Olson, P., & Pastalan, L.A. (1981). Effects of driver age on nighttime legibility of traffic signs. *Human Factors*, 23, 59–64.

Smiley, A. (1987). Effects of minor tranquilizers and antidepressants on psychomotor performance. *Journal of Clinical Psychiatry*, 48, 22–28.

Smiley, A. (2004). Adaptive strategies of older persons. In *Transportation in an Aging Society: A Decade of Experience*. Washington, DC: Transportation Research Board.

Smiley, A. (2006). *The BASICS: Functions Needed for Driving and Models of Driving*. Paper presented at the Transportation Injury Research Foundation and Office of the Superintendent of Motor Vehicles' Driving and Function Forum, Vancouver, BC.

Snoddy, G.S. (1926). Learning and stability. *Journal of Applied Psychology*, 10, 1–36.

Snook, K. (2008). Roles and responsibilities of licensing agencies. In D.W. Eby & L.J. Molnar (Eds.), *Proceedings of the North American License Policies Workshop*. Washington, DC: AAA Foundation for Traffic Safety.

Songer, T.J., LaPorte, R.E., Dorman, J.S., Orchard, T.J., Cruickshanks, K.J., Becker, D.J., & Drash, A.L. (1988). Motor vehicle accidents and IDDM. *Diabetes Care*, 11, 701–707.

Spain, D. (1997). *Societal Trends: The Aging Baby Boom and Women's Increased Independence*. Charlottesville, VA: University of Virginia.

Spector, A. (1982). Aging of the lens and cataract formation. In R. Sekular, D.W. Kline, & K. Dismukes (Eds.), *Aging and Human Visual Functions*. New York, NY: Alan Liss.

Spooner, J.W., Sakala, S.M., & Baloh, R.W. (1980). Effect of aging on eye tracking. *Archives of Neurology*, 37, 575–576.

Spreitzer-Berent, B. (1999). *Supporting the Mature Driver: A Handbook for Friends, Family Members, and Advisors*. Royal Oak, MI: AgeQuest.

Spudis, E.V., Penry, J.K., & Gibson, P. (1986). Driving impairment caused by episodic brain dysfunction: Restrictions for epilepsy and syncope. *Archives of Neurology*, 43, 558–564.

Stacey, B. & Kendig, H. (1997). Driving, cessation of driving, and transport safety issues among older people. *Health Promotion Journal of Australia*, 7, 175–179.

Ståhl, A., Oxley, P., Berntman, M., & Lind, L. (1994). The use of vision enhancements to assist elderly drivers. In *Towards an Intelligent Transport System: Proceedings of the First World Congress on Applications of Transport Telematics and Intelligent Vehicle-Highway Systems*. London, England: Artech House.

Stalvey, B.T. & Owsley, C. (2000). Self-perceptions and current practices of high-risk older drivers: Implications for driver safety interventions. *Journal of Health Psychology*, 5, 441–456.

Stalvey, B.T. & Owsley, C. (2003). The development and efficacy of a theory-based educational curriculum to promote self-regulation among high-risk older drivers. *Health Promotion Practice*, 4, 109–119.

Stamatiadis, N. (1998). ITS and human factors and the older driver: The U.S. experience. *Transportation Quarterly*, 52, 91–101.

Stamatiadis, N. (2001). Is ITS ready for the older driver? In *2001 IEEE Intelligent Transportation Systems Conference Proceedings*. New Brunswick, NJ: IEEE.

Stamatiadis, N., Taylor, W., & McKelvey, F. (1991). Elderly drivers and intersection accidents. *Transportation Quarterly*, 45, 377–390.

Stanton, N.A. & Young, M.S. (2005). Driver behaviour with adaptive cruise control. *Ergonomics*, 48, 1294–1313.

Staplin, L. (2008). Driver screening and assessment in the 21st century. In D.W. Eby & L.J. Molnar (Eds.), *Proceedings of the North American License Policies Workshop*. Washington, DC: AAA Foundation for Traffic Safety.

Staplin, L. & Dinh-Zarr, T.B. (2006). Promoting rehabilitation of safe driving abilities through computer-based clinical and personal screening techniques. *Topics in Geriatric Rehabilitation*, 22,129–138.

Staplin, L. & Freund, K. (2005). Public and private policy initiatives to move seniors forward. *Public Policy and Aging Report*, 15(2), 1–5.

Staplin, L., Gish, K., & Wagner, E. (2003). Mary PODS revisited: Updated crash analysis and implications for screening program implementation. *Journal of Safety Research*, 34, 389–397.

Staplin, L., Gish, K.W., & Joyce, J. (2008). "Low mileage bias" and related policy implications—A cautionary note. *Accident Analysis & Prevention*, 40, 1249–1252.

Staplin, L., Gish, K.W., Decina, L.E., Lococo, K.H., & McKnight, A.S. (1998). *Intersection Negotiation Problems of Older Drivers: Volume I. Final Technical Report*. Kulpsville, PA: The Scientex Corporation.

Staplin, L. & Lococo, K.H. (2003). *Model Driver Screening and Evaluation Program: Guidelines for Motor Vehicle Administrators* (Report No. DOT HS 809 581). Washington, DC: U.S. Department of Transportation.

Staplin, L., Lococo, K., Byington, S., & Harkey, D. (2001). *Guidelines and Recommendations to Accommodate Older Drivers and Pedestrians*. (Report No. FHWA-RD-01-051). Washington, DC: Federal Highway Administration.

Staplin, L., Lococo, K.H., Gish, K.W., & Decina, L.E. (2003a). *Model Driver Screening and Evaluation Program, Final Technical Report, Volume I: Project Summary and Model Program Recommendations*. (Report No. DOT HS 809 582). Washington, DC: U.S. Department of Transportation.

Staplin, L., Lococo, K.H., Gish, K.W., & Decina, L.E. (2003b). *Model Driver Screening and Evaluation Program, Final Technical Report, Volume II: Maryland Pilot Older Driver Study*. (Report No. DOT HS 809 583). Washington, DC: U.S. Department of Transportation.

Staplin, L., Lococo, K., & Sim, J. (1990). *Traffic Control Design Elements for Accommodating Drivers With Diminished Capacity: Volume II*. (Report No. FHWA-RD-90-055). Washington, DC: Federal Highway Administration.

Staplin, L., Lococo, K., Sim, J., & Drapcho, M. (1989). Age differences in visual information processing capability underlying traffic control device usage. *Transportation Research Record*, 1244, 63–72.

Staplin, L., Lococo, K.H., McKnight, A.J., McKnight, A.S., & Odenheimer, G.L. (1998). *Intersection Negotiation Problems of Older Drivers, Volume II: Background Synthesis on Age and Intersection Driving Difficulties*. (Report No. DOT HS 808 850). Washington, DC: U.S. Department of Transportation.

Staplin, L., Lococo, K.H., Stewart, J., & Decina, L. (1999). *Safe Mobility for Older People: Notebook* (Report No. DOT HS 808 853). Washington, DC: National Highway Traffic Safety Administration.

States, J.D. (1985). Musculo-skeletal system impairment related to safety and comfort of drivers 55+. In J.W. Malfetti (Ed.), *Proceedings of the Older Driver Colloquium*. Falls Church, VA: AAA Foundation for Traffic Safety.

Stav, W.B., Hunt, L.A., & Arbesman, M. (2006). *Occupational Therapy Practice Guidelines for Driving and Community Mobility for Older Adults*. Bethesda, MD: American Occupational Therapy Association.

Steel, J.M., Frier, N.M., Young, R.J., & Duncan, L.J.P. (1981). Driving and insulin-dependent diabetes. *The Lancet*, 318, 354–356.

Sternberg, S. (1966). High-speed scanning in human memory. *Science,* 153, 652–654.

Sterns, H.L., Sterns, R., Aizenberg, R., & Anapole, J. (2001). *Family and Friends Concerned About an Older Driver.* (Report No. DOT HS 809 307). Washington, DC: U.S. Department of Transportation.

Stevens, A.B., Roberts, M., McKane, R., Atkinsons, A.B., Bell, P.M., & Hayes, J.R. (1989). Motor vehicle driving among diabetics taking insulin and non-diabetics. *British Medical Journal,* 299, 591–595.

Stewart, R.B., Moore, M.T., Marks, R.G., May, F.E., & Hale, W.E. (1993). *Driving Cessation and Accidents in the Elderly: An Analysis of Symptoms, Diseases, Cognitive Dysfunction and Medication.* Washington, DC: AAA Foundation for Traffic Safety.

Stocker, F.W. & Moore, L.W. Jr. (1975). Detecting changes in the cornea that come with age. *Geriatrics,* 30, 57–69.

Stolwyk, R.J., Charlton, J.T., Triggs, T.J., Iansek, R., & Bradshaw, J.L. (2006). Neuropsychological function and driving ability in people with Parkinson's disease. *Journal of Clinical and Experimental Neuropsychology,* 28, 898–913.

Stolwyk, R.J., Triggs, T.J., Charlton, J.L., Moss, S., Iansek, R., & Bradshaw, J.L. (2006). Effect of a concurrent task on driving performance in people with Parkinson's disease. *Movement Disorders,* 21, 2096–2100.

Stone, J.R., Chae, K., & Pillalamarri, S. (2002). *The Effects of Roundabouts on Pedestrian Safety.* Raleigh, NC: North Carolina State University, Department of Civil Engineering.

Stuck, A.E., Beers, M.H., Steiner, A., Aronow, H.U., Rubenstein, L.Z., & Beck, J.C. (1994). Inappropriate medication use in community-residing older persons. *Archives of Internal Medicine,* 154, 2195–2200.

Stutts, J. (2005). *Improving the Safety of Older Road Users: A Synthesis of Highway Practice.* (NCHRP Synthesis 348). Washington, DC: Transportation Research Board.

Stutts, J.C., Reinfurt, D.W., & Rodgman, E.A. (2001). The role of driver distraction in crashes: An analysis of 1995–1999 Crashworthiness Data System data. In *45th Annual Proceedings of the Association for the Advancement of Automotive Medicine.* Des Plaines, IA: AAAM.

Stutts, J.C., Stewart, J.R., & Martell, C. (1998). Cognitive test performance and crash risk in an older driver population. *Accident Analysis & Prevention,* 30, 337–346.

Stutts, J.C., Wilkins, J.W., Reinfurt, D.W., Rodgman, E.A., & Van Heusen-Causey, S. (2001). *The Premature Reduction and Cessation of Driving by Older Men and Women.* Chapel Hill, NC: University of North Carolina Highway Safety Research Center.

Suen, L. & Mitchell, C.G.B. (1998). The value of intelligent transport systems to elderly and disabled travelers. In *Setting the Pace—Eighth International Conference on Transport and Mobility for Elderly and Disabled People, Volume 1.* Perth, Western Australia: Indomed Pty, Ltd.

Suen, S. L. & Sen, L. (2004). Mobility options for seniors. In *Transportation in an Ageing Society Conference Proceedings,* 27. Washington, DC: Transportation Research Board.

Sullivan, J.M., Bärgman, J., Adachi, G., & Schoettle, B. (2004). *Driver Performance and Workload Using a Night Vision System.* (Report No. UMTRI-2004-8). Ann Arbor, MI: University of Michigan Transportation Research Institute.

Summala, H. (1996). Accident risk and driver behaviour. *Safety Science,* 22, 103–117.

Sussman, E.D., Bishop, H., Madnick, B., & Walter, R. (1985). Driver inattention and highway safety. *Transportation Research Record,* 1047, 40–48.

Szafran, J. (1953). *Some Experiments on Motor Performance in Relation to Aging.* Unpublished thesis, Cambridge University.

Szlyk, J.P., Alexander, K.R., Severing, K., & Fishman, G.A. (1992). Assessment of driving performance in patients with retinitis pigmentosa. *Archives of Ophthalmology,* 110, 1709–1713.

Szlyk, J.P., Fishman, G.A., Severing, K., Alexander, K.R., & Viana, M. (1993). Evaluation of driving performance in patients with juvenile macular degeneration. *Archives of Ophthalmology*, 111, 207–212.

Szlyk, J.P., Mahler, C.L., Seiple, W., Edward, D.P., & Wilensky, J.T. (2005). Driving performance of glaucoma patients correlates with peripheral visual vision loss. *Journal of Glaucoma*, 14, 145–150.

Szlyk, J.P., Mahler, C.L., Seiple, W., Vajaranant, T.S., Blair, N.P., & Shahidi, M. (2004). Relationship of retinal structural and clinical vision parameters to driving performance of diabetic retinopathy patients. *Journal of Rehabilitation Research & Development*, 41, 347–358.

Szlyk, J.P., Myers, L., Zhang, Y., Wetzel, L., & Shapiro, R. (2002). Development and assessment of a neuropsychological battery to aid in predicting driving performance. *Journal of Rehabilitation and Development*, 39, 483–496.

Szlyk, J.P., Pizzimenti, C.E., Fishman, G.A., Kelsch, R., Wetzel, L.C., Kagan, S., & Ho, K. (1995). A comparison of driving in older subjects with and without age-related macular degeneration. *Archives of Ophthalmology*, 113, 1033–1040.

Szlyk, J.P., Seiple, W., & Viana, M. (1995). Relative effects of age and compromised vision on driving performance. *Human Factors*, 37, 430.

Szlyk, J.P., Taglia, D.P., Paliga, J., Edward, D.P., & Wilensky, J.T. (2002). Driving performance in patients with mild to moderate glaucomatous clinical vision changes. *Journal of Rehabilitation Research & Development*, 39, 467–481.

Taub, H.A. (1974). Coding for short-term memory as a function of age. *Journal of Genetic Psychology*, 125, 309–314.

Taylor, D.H. (1964). Drivers' galvanic skin response and the risk of accident. *Ergonomics*, 7, 439–451.

Taylor, J., Chadwick, D.W., & Johnson, T. (1995). Accident experience and notification rates in people with recent seizures, epilepsy or undiagnosed episodes of loss of consciousness. *Quarterly Journal of Medicine*, 88, 733–740.

Terán-Santos, J., Jiménez-Gómez, A., & Cordero-Guevara, J. (1999). The association between sleep apnea and the risk of traffic accidents. *New England Journal of Medicine*, 340, 847–851.

Terry, R.D. & Katzman, R. (1983). Senile dementia of the Alzheimer type. *Annals of Neurology*, 14, 497–506.

Thomas, J.C., Waugh, N.C., & Fozard, J.L. (1978). Age and familiarity in memory scanning. *Journal of Gerontology*, 33, 528–533.

Torpey, S.E. (1986). *License Re-Testing of Older Drivers*. Melbourne, AU: Road Safety Authority.

Traffic Injury Research Foundation (2006). *Driving and Function Forum*. URL: http://www.trafficinjuryresearch.com/driving_and_function/forum_overview.htm.

Transportation Research Board. (2004). *Proceedings of Transportation in an Aging Society: A Decade of Experience*. Washington, DC: National Academy of Sciences.

Trick, G.L. & Silverman, S.E. (1991). Visual sensitivity to motion: Age-related changes and deficits in senile dementia of the Alzheimer type. *Neurology*, 41, 1437–1440.

Triesman, A.M. & Gelade, G. (1980). A feature integration theory of attention. *Cognitive Psychology*, 12, 97–136.

Trobe, J.D., Waller, P.F., Cook-Flannagan, C.A., Teshima, S.M., & Bieliauskas, L.A. (1996). Crashes and violations among drivers with Alzheimer disease. *Archives of Neurology*, 53, 411–416.

Tulving, E. (1974). Recall and recognition of semantically encoded words. *Journal of Experimental Psychology*, 102, 778–787.

Tuokko, H., Beattie, L., Tallman, K., & Cooper, P. (1995). Predictors of motor vehicle crashes in a dementia clinic population: The role of gender and arthritis. *Journal of the American Geriatrics Society*, 43, 1444–1445.

Uc, E.Y., Rizzo, M., Anderson, S.W., Shi, Q., & Dawson, J.D. (2005). Driver landmark and traffic sign identification in early Alzheimer's disease. *Journal of Neurology, Neurosurgery & Psychiatry*, 76, 764–768.

Uc, E.Y., Rizzo, M., Anderson, S.W., Sparks, J., Rodnitzky, R.L., & Dawson, J.D. (2006a). Impaired visual search in drivers with Parkinson's disease. *Annuals of Neurology*, 60, 407–413.

Uc, E.Y., Rizzo, M., Anderson, S.W., Sparks, J., Rodnitzky, R.L., & Dawson, J.D. (2006b). Driving with distractions in Parkinson's disease. *Neurology*, 67, 1774–1780.

Ulf, B. & Jorgen, L. (1999). *Traffic Safety of Roundabouts for Cyclists and Pedestrians*. Linkoping, Sweden: Swedish National Road and Transport Research Institute.

Ullman, G.L. (1993). Motorist interpretation of MUTCD freeway lane control signals. *Transportation Research Record*, 1403, 49–56.

Underwood, M. (1992). The older driver: Clinical assessment and injury prevention. *Archives of Internal Medicine*, 152, 735–740.

University of Louisiana at Monroe School of Nursing (2008). *Nursing Teaching and Learning Principles*. URL: http://www.ulm.edu/nursing/learningprinciples.doc .

Unsworth, C.A., Wells, Y., Browning, C., Thomas, S.A., & Kendig, H. (2007). To continue, modify or relinquish driving: Findings from a longitudinal study of healthy aging. *Gerontology*, 53, 423–431.

U.S. Census Bureau. (2004). *U.S. Interim Projections by Age, Sex, Race, and Hispanic Origin*. http://www.census.gov/ipc/www/usinterimproj/ Released March 18, 2004.

U.S. Census Bureau. (2006). *Oldest Baby Boomers Turn 60!* http://www.census.gov/Press-Release/www/releases/archives/facts_for_features_special_editions/006105.html. Released January 3, 2006.

U.S. Census Bureau. (2008). *Births, Deaths, Marriages, and Divorce: Life Expectancy* http://www.census.gov/compendia/statab/cats/births_deaths_marriages_divorces/life_expectancy.html. Released March 17, 2008.

U.S. Department of Health and Human Services. (1999). *Mental Health: A Report of the Surgeon General*. Rockville, MD: U.S. Department of Health and Human Services, National Institute of Mental Health.

U.S. Department of Health and Human Services. (2005). *Seniors Benefit from Transportation Coordination Partnerships—a Toolbox*. Washington DC: U.S. Department of Health and Human Services.

Vanderpump, M.P., Tunbridge, W.M., French, J.M., Appleton, D., Bates, D., Clark, F., Grimley, E.J., Hasan, D.M., Rodgers, H., Tunbridge, F., & Young, E.T. (1995). The incidence of thyroid disorders in the community: A twenty-year follow-up of the Whickham Survey. *Clinical Endocrinology*, 43, 55–68.

Van Wolffelaar, P. & Rothengatter, T. (1990). Divided attention in RTI-tasks for elderly drivers. *EC DRIVE Programme, Project V1006: DRIVAGE*. Groningen, The Netherlands: Traffic Research Center, University of Groningen.

Van Wolffelaar, P.C., Brouwer, W.H., & Rothengatter, J.A. (1991). Older drivers handling road traffic informatics: Divided attention in a dynamic driving simulator. *Proceedings of the Conference: Strategic Highway Research Program and Traffic Safety on Two Continents*. Gothenburg, Sweden.

Van Wolffelaar, P.C., Rothengatter, T., & Brouwer, W. (1991). Elderly drivers' traffic merging decisions. In A.G. Gale et al. (Eds.), *Vision in Vehicles III*. Amsterdam, The Netherlands: Elsevier.

Vernon, D.D., Diller, E.M., Cook, L.J., Reading, J.C., & Deane, J.M. (2001). *Further Analysis of Drivers Licensed With Medical Conditions in Utah*. (Report No. DOT HS 809 211). Washington, DC: U.S. Department of Transportation.

Verriest, G. (1971). L'influence de l'age sur les fonctions visuelles de l'homme. [in French]. *Bull de l'Académie Royale de Mèdecine Belgique*, 11, 264–265.

Verster, J.C. & Volkerts, E. R. (2004). Antihistamines and driving ability: Evidence from on-the-road driving studies during normal traffic. *Annals of Allergy, Asthma, & Immunology*, 92, 294–304.

Vingrys, A.J. & Cole, B.L. (1988). Are colour vision standards justified in the transport industry? *Ophthalmic & Physiological Optics*, 8, 257–274.

Vivano, D.C., Culver, C.C., Evans, L., Frick, M., & Scott, R. (1989). Involvement of older drivers in multi-vehicle side impact crashes. *Proceedings of the 12th International Technical Conference of Experimental Safety Vehicles*. Washington, DC: National Highway Traffic Safety Administration.

Vivano, D.C., Culver, C.C., Evans, L., Frick, M., & Scott, R. (1990). Involvement of older drivers in multiple-vehicle side impact crashes. *Accident Analysis & Prevention*, 22, 178–188.

Vrkljan, B.H. & Polgar, J.M. (2007). Driving, navigation, and vehicular technology: Experiences of older drivers and their co-pilots. *Traffic Injury Prevention*, 8, 403–410.

Wacker, J., Busser, A., & Lachenmayr, J. (1983). Influence of stimulus size and contrast on the temporal patterns of saccadic eye movements: Implications for road traffic. *German Journal of Ophthalmology*, 2, 246–250.

Waddell, E. (2008). *Evolution of Roundabout Technology: A History-Based Literature Review*. URL: http://www.k-state.edu/roundabouts/research/Waddell.pdf. Accessed March, 2008.

Wall, R., Long, R., Guth, D., Ashmead, D., & Ponchillia, P. (2005). Roundabouts: Problems of and strategies for success. *International Congress Series*, 1282, 1085–1088.

Wallach, H. & O'Connell, D.N. (1953). The kinetic depth effect. *Journal of Experimental Psychology*, 45, 205–217.

Waller, J.A. (1965). Chronic medical conditions and traffic safety. *The New England Journal of Medicine*, 273, 1413–1420.

Walsh, D.A., Krauss, I.K., & Regnier, V.A. (1981). Spatial ability, environmental knowledge, and environmental use: The elderly. In L. Libon, A. Patterson, & N.I. Newcombe (Eds.), *Spatial Representation and Behavior Across the Life Span*. New York, NY: Academic Press.

Walsh, J.M., De Gier, J.J., Christopherson, A.S., & Verstraete, A.G. (2004). Drugs and driving. *Traffic Injury Prevention*, 5, 241–253.

Wang, C.C., Kosinski, C.J., Schwartzberg, J.G., & Shanklin, A.V. (2003). *Physician's Guide to Assessing and Counseling Older Drivers*. Chicago, IL: American Medical Association.

Wang, J.-S., Knipling, R.R., & Goodman, M.J. (1996). The role of driver inattention in crashes: New statistics from the 1995 Crashworthiness Data System. In *40th Annual Proceedings Association for the Advancement of Automotive Medicine*. Des Plaines, IA: AAAM.

Warabi, T., Kase, M., & Kato, T. (1984). Effect of aging on the accuracy of visually guided saccadic eye movement. *Annals of Neurology*, 16, 449–454.

Warlow, C. & Morris, P.J. (1982). *Transient Ischemic Attacks*. New York, NY: M. Dekker.

Weale, R.A. (1971). The ageing eye. In *Scientific Basis of Medicine: Annual Reviews*. London, UK: Athlone Press.

Weale, R.A. (1982). *A Biography of the Eye*. London, UK: HK Lewis.

WebMD. (2006). *Depression Basics*. URL: http://www.webmd.com/depression/guide/depression-basics.

WebMD. (2008). *Understanding Low Blood Pressure: The Basics*. URL: http://www.webmd.com/hypertension-high-blood-pressure/guide/understanding-low-blood-pressure-basics.

Weir, D.H. & McRuer, D.T. (1968). A theory for driver steering control of motor vehicles. *Highway Research Record*, 247, 7–39.

Welford, A.T. (1959). Psychomotor performance. *Annual Review of Gerontology & Geriatrics*, 4, 237.

West, R.L., Crook, T.H., & Barron, K.L. (1992). Everyday memory performance across the life-span: Effects of age and noncognitive individual differences. *Psychology & Aging,* 7, 72–82.

Westchester County Department of Senior Programs and Services. (n.d.). *Older Driver Family Assistance Program: When It's Time to Have "The Talk" With an Older Driver.* Yorktown Heights, NY: Westchester County Department of Senior Programs and Services.

Westheimer, G. (1986). The eye as an optical instrument. In K.R. Boff, L. Kaufman, & J.P. Thomas (Eds.), *Handbook of Perception and Human Performance: Vol. I. Sensory Processes and Perception.* New York, NY: John Wiley and Sons.

Westlake, W. (2001). Is a one eyed racing driver safe to compete? Formula one (eye) or two? *British Journal of Ophthalmology,* 85, 619–624.

Wheatley, C.J., Pellerito, J.M., Jr., & Redpenning, S. (2006). The clinical evaluation. In J.M. Pellerito (Ed.), *Driver Rehabilitation and Community Mobility: Principles and Practice.* St. Louis, MO: Elsevier Mosby.

Whelan, M., Langford, J., Oxley, J., Koppel, S., & Charlton, J. (2006). *The Elderly and Mobility: A Review of the Literature (Report No. 255).* Victoria, Australia: Monash University Accident Research Centre.

Whelihan, W.M., DiCarlo, M.A., & Paul, R.H. (2005). The relationship of neuropsychological functioning to driving competence in older persons with early cognitive decline. *Archives of Clinical Neuropsychology,* 20, 217–228.

White House Conference on Aging. (2005). *Booming Dynamics of Aging: From Awareness to Action: Final Report.* Washington, DC: White Conference on Aging.

Wilde, G.J. (1982). The theory of risk homeostasis: Implications for safety and health. *Risk Analysis,* 2, 209–225.

Wilde, G.J. & Murdoch, P.A. (1982). Incentive systems for accident-free and violation-free driving in the general population. *Ergonomics,* 25, 879–890.

Wilkinson, C.J. & Moskowitz, H. (2001). *Polypharmacy & Older Drivers—Literature Review.* (unpublished document). Los Angeles, CA: Southern California Research Institute.

Williams, L.E. (2002). *Emergency Vehicle Automatic Crash Notification & Event Reporting Technology.* Aliso Viejo, CA: Roadside Telematics Corporation.

Williams, S.A., Denney, N.W., & Schadler, M. (1983). Elderly adults' perception of their own cognitive development during adult years. *International Journal of Aging & Human Development,* 16, 147–158.

Wilson, M.R. (1989). Glaucoma in Blacks: Where do we go from here? *Journal of the American Medical Association,* 261, 281–282.

Windsor, T.D., Anstey, K.J., Butterworth, P., Luszcz, M.A., & Andrews, G.R. (2007). The role of perceived control in explaining depressive symptoms associated with driving cessation in a longitudinal study. *The Gerontologist,* 47, 215–223.

Winter Park Health Foundation. (2006). *Florida's Volunteers: The Driving Force for Senior Mobility.* Winter Park, FL: Winter Park Health Foundation. URL: www.unitedweride. gov/Volunteer_Drivers_for_Senior_Transportation_12-2006.pdf

Wochinger, K. & Boehm-Davis, D. (1995). *The Effects of Age, Spatial Ability, and Navigation Information on Navigational Performance.* (Report No. FHWA-RD-95-166). Washington, DC: Federal Highway Administration.

Wolf, E. (1960). Glare and age. *Archives of Ophthalmology,* 64, 502–514.

Wolf, E. (1967). Studies on the shrinkage of the visual field with age. *Transportation Research Record,* 164, 1–7.

Wood, J.M. & Troutbeck, R. (1994). Effect of visual impairment on driving. *Human Factors,* 36, 476–487.

Wood, J.M. & Troutbeck, R. (1995). Elderly drivers and simulated visual impairment. *Optometry and Vision Science,* 72, 115–124.

Wood, J.M., Worringham, C., Kerr, G., & Silburn, P. (2004). Quantitative assessment of driving performance in Parkinson's disease. *Journal of Neurology, Neurosurgery, & Psychiatry*, 76, 176–180.

Workzonesafety.org. (2008). *Work Zone Fatalities: 2006*. URL: http://www.workzonesafety.org/crash_data/.

Wright, B.M. & Paine, R.B. (1985). Effects of aging on sex differences in psychomotor reminiscence and tracking proficiency. *Journal of Gerontology*, 40, 179–184.

Wright, R.A. (1996). *A Brief History of the First 100 Years of the Automobile Industry in the United States*. Detroit, MI: Wayne State University Department of Communication.

Wright, R.E. (1981). Aging, divided attention, and processing capacity. *Journal of Gerontology*, 37, 76–79.

Yee, D. (1985). A survey of the traffic safety needs and problems of drivers age 55 and over. In J.W. Malfetti (Ed.), *Needs and Problems of Older Drivers: Survey Results and Recommendations*. Falls Church, VA: AAA Foundation for Traffic Safety.

Young, T., Blustein, J., Finn, L., & Palta, M. (1997). Sleep-disordered breathing and motor vehicle accidents in a population-based sample of employed adults. *Sleep*, 20, 608–613.

Zein, S. & Mairs, A. (2002). *AAA Michigan Road Improvement Demonstration Program Evaluation*. British Columbia, Canada: G.D. Hamilton Associates.

Zesiewicz, T.A., Cimino, C.R., Malek, A.R., Gardner, N., Leaverton, P.L., & Dunne, P.B. (2002). Driving safety in Parkinson's disease. *Neurology*, 59, 1787–1788.

Zhan, C., Sangl, J., Bierman, A.S., Miller, M.R., Friedman, B., Wickizer, S.W., & Meyer, G.S. (2001). Potentially inappropriate medication use in the community-dwelling elderly. *Journal of the American Medical Association*, 286, 2832–2829.

Zhang, J., Lindsay, J., Clarke, K., Robbins, G., & Mao, Y. (2000). Factors affecting the severity of motor vehicle traffic crashes involving elderly drivers in Ontario. *Accident Analysis & Prevention*, 32, 117–125.

Zhao, C., Popovic, V., & Ferreira, L. (2007). Elderly drivers' needs: Methodological approach and its application to the educational context. In *Proceedings Connected 2007: International Conference on Design Education*. Sydney, Australia: University of New South Wales.

Zhao, C., Popovic, V., Ferreira, L., & Lu, X. (2006). Vehicle design research for Chinese elderly drivers. *Proceedings Computer-Aided Industrial Design and Conceptual Design, 2006*. New Brunswick, NJ: IEEE.

Zwahlen, H.T. & Schnell, T. (1999). Visibility of road marking as a function of age, retroreflectivity under low-beam and high-beam illumination at night. *Transportation Research Record*, 1692, 152–162.

Index

Milton Keynes UK
Ingram Content Group UK Ltd.
UKHW040104071024
449327UK00019B/807